Algebra Ready

YOU CAN BE ALGEBRA READY

Ken Andrews, M.A.
Diane Johnson, M.A.

PART I

TEACHER'S GUIDE

BOSTON, MA • NEW YORK, NY • LONGMONT, CO

Copyright 2003 by Sopris West Educational Services
All rights reserved.

No part of this work may be reproduced or transmitted in any form or by any means, electronic or mechanical, including photocopying or recording, or by any information retrieval system, without the express written permission of the publisher.

07 06 05   5 4 3

ISBN 1-57035-846-X

Printed in the United States of America

Published and Distributed by

SOPRIS WEST
EDUCATIONAL SERVICES

4093 Specialty Place • Longmont, CO 80504 • (800) 547-6747
www.sopriswest.com

185TE1/18-05

Acknowledgments

This work would not have been possible without the interest of many practitioners in the field of mathematics. We offer our sincere thanks to everyone who helped to make this work a reality. In particular, we thank the Denver Public Schools for their proactive interest in supporting the creation of a mathematics program designed to meet the needs of students of all ability levels and also for making available to the authors the many human and physical resources of the school district.

A special thank you to Jeanne Jessup, a practicing middle school mathematics teacher in a large urban school district, who used the materials in her math classes over the past several years. She provided valuable feedback regarding classroom practices, student acquisition and application of concepts, and student reactions to the innovative teaching strategies. In addition, Jeanne offered her skills as a mathematician to serve as a final editor in preparing *Algebra Ready* for publication.

We also gratefully acknowledge Kate Gallaway, M.A., Educational Therapist, San Rafael, California, for her comments and expertise in the area of mathematics and special education and Joe Witt, Ph.D., Louisiana State University, for his contributions relating to educational assessment and management. We appreciate their thoughtful reviews and commonsense feedback regarding the intent of *Algebra Ready,* its content, format, and compatibility with accepted instructional practices, and its impact on the math achievement of students of all ability levels, both in the general education classroom and in alternative settings.

About the Authors

Ken Andrews

During his career in education, Ken Andrews has held the positions of teacher; principal; psychologist; curriculum and assessment developer; director of testing and evaluation; director of planning, research, and program evaluation; and, most recently, author, educational consultant and researcher in the areas of teaching, learning, and increasing student achievement. He has devoted most of his professional life to learning theory and its application to efficient classroom instruction, and his contributions to educational research have resulted in the development of many innovative teaching methodologies, learning systems, and assessment formats and technologies. In addition to his work in the schools, he was founder and director of the nationally known Denver Diagnostic Teaching Centers, a unique model that has served as a guideline for similar centers throughout the country, and cofounder, with coauthor Diane Johnson, of the Learner's Edge, Inc., described below.

Diane Johnson

Diane Johnson is an accomplished teacher, principal, educational trainer, and author. She is a specialist in the areas of curriculum, instruction, and assessment development and implementation. During her career as an educator, she worked with students and adults of all ages and ability levels in all basic academic subjects. Her varied background includes living and working in Mexico and southeast Asia as well as throughout the United States. Most recently she cofounded, with Ken Andrews, The Learner's Edge, Inc., an educational consulting firm devoted to the research and development of achievement-oriented learning programs and materials.

Established in 1998, the Learner's Edge, Inc., is an educational consulting/publishing firm devoted to making quality educational materials and services available to the education community and the public. In addition to *Algebra Ready,* the firm has created more than a dozen successful teaching/learning programs, including *Little Words for Little People,* a beginning reading program for grades K–2; *Jump Start,* a series of reading programs for grade 3 through adulthood; *Think Sheets,* a reading comprehension/critical-thinking program; and *Reading Achievement at a Glance* and *Math Skills at a Glance,* standards-based performance assessments that sample typical skills and question formats found on most state-standards assessments.

Contents

Part 1

Introduction	1
You Can Be Algebra Ready	3
Lesson Components	5
Lesson 1	7

Perform the order of operations: × ÷ and + − only.
Practice addition, subtraction, multiplication, and division.

Lesson 2	15

Perform the order of operations: exponents, × ÷, and + −.
Practice addition, subtraction, multiplication, and division.

Lesson 3	19

Perform the order of operations: parentheses and other grouping symbols (brackets and braces), exponents, × ÷, and + −.

Lesson 4	25

Review the order of operations: parentheses and other grouping symbols (brackets and braces), exponents, × ÷, and + −.
Please **E**xcuse **M**y **D**ear **A**unt **S**ally

Try This	33
Lesson 5	35

Evaluate formulas, specifically the area of a circle. $A = \pi r^2$

Lesson 6	41

Combine integers using the number line.
Apply the first three signed number rules.

Lesson 7	47

Apply signed number rules 1 through 6.
Review signed number rules 1 through 3.

Lesson 8	53

Apply the six signed number rules.
Review the order of operations.

Lesson 9	59

Apply the six signed number rules including eliminating a double sign.
Review the order of operations.

Lesson 10	65

Review the order of operations using several types of grouping symbols.
Practice the six signed number rules.

Try This	71
Lesson 11	73

Review the order of operations and the signed number rules.
Practice addition, subtraction, multiplication, division, and exponents.

Lesson 12	81

Solve addition and subtraction equations with variables.

Try This	87
Lesson 13	89

    Solve multiplication equations with variables.
    Review addition and subtraction equations.
    Convert improper fractions.

Lesson 14	97

    Solve two-step equations.
    Reduce fractions.
    Practice addition, subtraction, multiplication, and division with integers.

Lesson 15	105

    Review two-step equations.
    Practice addition, subtraction, multiplication, and division with integers
    and fractions.

Lesson 16	113

    Simplify expressions by combining like terms.
    Add and subtract with integers.

Lesson 17	119

    Simplify expressions by combining like terms.
    Add and subtract with integers.

Lesson 18	125

    Solve two-step equations with variable terms on both sides.
    Practice the four basic operations with integers and fractions.

Try This	133
Resources	**135**
Bookmark	
Explanation: Order of Operations	137
Vocabulary Words by Lesson	139
Diagnostic Assessment—Subtest 1	143
Diagnostic Assessment—Subtest 2	147
Diagnostic Assessment—Subtest 3	151
Diagnostic Assessment—Subtest 4	155
Sample Diagnostic Assessment Results by Student and by Class	159
Diagnostic Assessment Report—Subtest 1	161
Diagnostic Assessment Report—Subtest 2	162
Diagnostic Assessment Report—Subtest 3	163
Diagnostic Assessment Report—Subtest 4	164
Skill Practice	165
Additional Word Problem Warm-Ups	205
Glossary	211
Scope and Sequence	231

Algebra Ready Vocabulary Card Holder Pattern

Algebra Ready Student Completion Award

Lessons Included in Part 2

Introduction 1

You Can Be Algebra Ready 3

Lesson Components 5

Lesson 19 7
    Solve multi-step equations with variable terms on both sides.
    Practice the four basic operations with integers.

Lesson 20 15
    Review multi-step equations with variable terms on both sides.
    Practice the four basic operations with integers and fractions.

Lesson 21 23
    Evaluate formulas for geometric figures using diagrams and substitution.
    Multiply and divide whole numbers, fractions, and decimals.

Lesson 22 29
    Review two-step and multi-step equations.
    Practice the four basic operations with integers and fractions.

Try This 37

Lesson 23 39
    Solve inequalities.

Lesson 24 47
    Solve equations that have fractions.
    Practice working with fractions.

Lesson 25 55
    Solve equations with decimals.
    Practice the four basic operations with decimals.

Lesson 26 61
    Practice solving fraction and decimal equations.

Lesson 27 67
    Evaluate formulas for geometric figures using data from diagrams.

Lesson 28 73
    Simplify monomials using multiplication.

Lesson 29 81
    Simplify algebraic expressions using the distributive property.
    Use the six signed number rules to simplify algebraic expressions.

Try This 89

Lesson 30 91
    Simplify more complex algebraic expressions using the distributive property.
    Apply the six rules of signed numbers to simplify algebraic expressions.

Lesson 31 97
    Simplify algebraic expressions using the distributive property and the
    signed number rules to distribute a plus (+) sign through parentheses.

Lesson 32 103
    Simplify algebraic expressions using the distributive property and the
    signed number rules to distribute a minus (−) sign through parentheses.

Lesson 33 109
    Solve equations using the distributive property and combining like terms.

Lesson 34 115
    Solve multi-step equations with variable terms on both sides.
    Practice using the order of operations, the signed number rules,
    the distributive property, and inverse operations.

Lesson 35 123
    Practice solving multi-step equations with variable terms on both sides.
    Practice using the order of operations, the signed number rules,
    the distributive property, and inverse operations.

Lesson 36	131
Apply algebra to real-world problem solving.	
Try This	143
Resources	**145**
Bookmark	
Explanation: Order of Operations	147
Vocabulary Words by Lesson	149
Diagnostic Assessment—Subtest 1	153
Diagnostic Assessment—Subtest 2	157
Diagnostic Assessment—Subtest 3	161
Diagnostic Assessment—Subtest 4	165
Sample Diagnostic Assessment Results by Student and by Class	169
Diagnostic Assessment Report—Subtest 1	171
Diagnostic Assessment Report—Subtest 2	172
Diagnostic Assessment Report—Subtest 3	173
Diagnostic Assessment Report—Subtest 4	174
Skill Practice	175
Additional Word Problem Warm-Ups	215
Glossary	221
Scope and Sequence	241
Algebra Ready Vocabulary Card Holder Pattern	
Algebra Ready Student Completion Award	

Introduction

Algebra Ready provides students with the essential math knowledge and experiences required to be successful in algebra and geometry. This carefully designed math program integrates the fundamentals of psychologically sound learning practices with the mathematical skills necessary for moving on to higher math. A comprehensive program, *Algebra Ready* provides students with direct instruction and the opportunity for both guided learning and independent practice.

Algebra Ready blends the teaching of fundamental math skills with instruction in the thinking processes required for success in entry-level algebra and geometry. The initial phases of *Algebra Ready* emphasize mastery of basic math. As students progress through the spiraling content of this program, they learn more advanced skills that are essential to higher mathematics. When students have satisfactorily completed the program, they have mastered fundamental mathematics and gained a working knowledge of the key concepts, skills, and strategies necessary for a successful transition to algebra and geometry.

The instructional format of *Algebra Ready* is based upon a supportive relationship between direct teacher instruction and written application. *Algebra Ready* uses a multisensory, kinesthetic approach that meets the needs of all students. All objectives are action oriented. Students are actively doing things; they are actively involved in learning. All lessons involve listening, observing, operating, and practicing.

Although this program is designed to be as student friendly as possible, best success is achieved when students work with an energetic, stimulating, and involved teacher. *Algebra Ready* teachers possess a thorough knowledge of mathematics and the instructional dynamics of the *Algebra Ready* program. They are spontaneous, flexible, and able to analyze student learning, making the necessary changes in delivery to ensure student success. *Algebra Ready* teachers are encouraged to be creative, using strategies that result in observable changes in student progress. A teacher's success with *Algebra Ready* is directly proportional to the degree of each teacher's ability, creativity, interest, and desire to excel.

Algebra Ready teachers are active participants in instruction. They interact extensively with their students throughout the lessons, guiding student learning and modeling appropriate skills, attitudes, and practices. The teachers conscientiously direct students' learning, helping them to focus on experiences, rules, procedures, and facts that build mathematical confidence.

Lessons are graded daily and quizzes are given regularly, allowing the teacher to closely monitor student progress. Class work, homework, and word problem warm-ups are designed to foster critical thinking and problem-solving skills. *Algebra Ready* activities encourage all students to make the program content their own. Students are engaged in learning through a variety of "doing" experiences and continually demonstrate their newly acquired expertise.

Algebra Ready's daily practice, hands-on teaching, constructive feedback, and continuous validation of success dispel feelings of math anxiety that prevent many students from fulfilling their potential. *Algebra Ready* develops self-confidence, and students respond to math with an "I can do it" attitude.

Algebra Ready includes a comprehensive diagnostic assessment that reflects the skills taught in each lesson and pinpoints those areas requiring instructional intervention or acceleration. This assessment may also be used to identify those students who already have the requisite skills for success in algebra and geometry and therefore may opt out of the program. When the assessment is used as a pretest, it determines present level of functioning and may serve as a source of baseline data. It can also be used as an ongoing assessment to determine the progress of an individual student or the class, so that teachers may fine-tune or review identified lessons. As a pre- and posttest, the assessment provides a beginning and an end to this course of study and determines individual student and class accomplishments.

Students who participate in *Algebra Ready* will have an advantage over many of their peers. They will enter algebra and geometry with a set of well-established skills and applications. They will have a confident "I can do it" attitude about their ability to be successful in higher mathematics, and they will have the skills to see that confidence through to reality.

You Can Be Algebra Ready

Read and discuss with your students.

Math is a skill that you use all the time. When you buy something at the store, you have to know how much money it will cost. You have to be able to tell if you have enough money and determine how much change you should receive. When you eat in a restaurant, you have to be able to figure out how much of a tip to leave. When something at a store is marked down, you have to be able to figure out what the new price is. What if you are planning a long car trip? You may need to figure out how many miles you can travel in a day, how many times you will need to buy gasoline, or how many days it will take you to cover the distance.

Plan ahead!

Every time you plan ahead to make enough money to buy something, you have to be able to figure out how much money you will need and how long it will take you to earn it. What if you want to buy a new car or a house? You may have to apply for a loan. You will have to figure out how much money you can afford to borrow and how much you can afford to pay back every month. Math is very important because you need it every day. You need to be confident with math if you want to have control of your life.

Did you ever think of math like you think of sports, art, or music? Math is exactly like those activities. Like them, math is a skill. Being confident at math takes time and practice and good teaching. Just like becoming an excellent artist, athlete, or musician requires a good coach, becoming a mathematician requires the expert guidance of a quality teacher. There are certain steps and rules in math that you must know and understand to be successful. You cannot just figure it out for yourself. You need to *learn* these rules and procedures from an expert, your *Algebra Ready* teacher. The next step is *practice*. At the beginning, math might seem confusing. Keep practicing, and it will become easier and easier. Eventually you will know how to do the work without a second thought. You will be amazed at how fast your math skills improve.

Practice, practice, practice!

Attitude and self-confidence make the difference.

Math matters!

As your math skills improve, your whole attitude about math and learning will improve. You will find yourself teaching other people how to do math. You may even find yourself helping your parents with the math that goes along with running a household and raising a family. You will feel like you can make plans about your future and be confident that they will work. You can set goals to buy things, save for vacations, save for college, and know that you can achieve your goals. Being more skilled and at home with math can open a whole new world of possibilities for career choices. Studies show that people who pass algebra do much better in college, are more optimistic about their future, and make more money during their lives. Math matters!

Commitment!

This program has been specifically designed to help you build your math skills, and as your math skills improve, your confidence will skyrocket. This program can't do it all for you, though. You have to make a commitment to learning. That means that you have to be ready to learn math. Your success will depend on your commitment to regular, steady practice with a determined effort to learn. You can do it! Think positively!

Do the work and it will tell you what to do!

Remember, common everyday math has become second nature to you. The same thing will happen as you complete this program. By doing the work regularly, with a determined attitude, you will learn it. You will master it. You will become a capable mathematician. *You will be algebra ready!*

Lesson Components

Algebra Ready lessons are composed of eight distinct parts: Word Problem Warm-Up, Lesson Objectives, Vocabulary Acquisition, Lesson Presentation, Guided Learning, Guided Class Practice, Independent Practice, and Try This. Each part is described below along with a recommended method for instruction. Individual teacher judgment will be required for pacing and fine-tuning the lessons to meet the specific needs of each class.

1. Word Problem Warm-Up

Word Problem Warm-Ups are activities designed to *maintain* and *reinforce* a variety of general math skills. They offer an opportunity for stimulating discussion, creativity, and the application of problem-solving skills. Teachers engage the students in a complete dialogue about each Word Problem Warm-Up.

Included with the Word Problem Warm-Up is a **problem-solving plan**, which presents students with a blueprint, or checklist, for solving problems. Students identify the steps and thinking processes used to arrive at their answer. This diagnostic tool gives teachers a quick, reliable indicator of how a student thinks and manipulates information required to solve a word problem. The plan is flexible and adaptive to student choice. Students indicate the steps taken in the order that they occurred by entering sequential numbers (1, 2, 3…). Where a step is repeated, the number for that step is entered in the second box.

These problem-solving steps (Read, Key Words, Analyze, Add, Subtract, Multiply, Divide, Compare/Check) are introduced to the students in the first lesson and are referred to as appropriate throughout the program or until the steps become an integrated part of the students' problem-solving skills.

Additional Word Problem Warm-Ups are provided in the Resources section of this handbook to be used as needed.

2. Lesson Objectives

Lesson Objectives are used by teachers to develop a mind-set for students and to set a direction for each lesson. These action-oriented objectives serve as a road map, showing students where they need to go and how to get there. As the tour guide, the teacher identifies and explains significant ideas and concepts along the way.

3. Vocabulary Acquisition

Common usage, high-frequency math words are printed on pocket-size flashcards and keyed to the lessons in which the words are introduced. Each student receives a set of cards to use and to keep. Not all of the vocabulary words are pertinent to the particular lesson. Those that are used in the lesson appear in italics in the list on the first page of the lesson. The other words appear in future lessons, are presented as review, or offer an opportunity to maintain common math knowledge. The degree of focus on nonitalicized words is left to teacher judgment.

On the front of each card, students are encouraged to draw a picture or symbol that, to them, illustrates the word or the definition or is an example of its usage. This emblem serves as a mnemonic device and helps the student to quickly remember the word. Teachers are encouraged to reinforce the vocabulary by using the cards in games and team contests, such as math vocabulary games that involve definitions, application, and conceptualization.

4. Lesson Presentation

At the start of each lesson, teachers review and relate previous learning to the lesson. The teacher then describes each new skill or concept orally, step-by-step, while at the same time presenting it visually. The use of the chalkboard or other visuals, as well as hands-on experiences or manipulatives where appropriate, is encouraged.

5. Guided Learning

During Guided Learning, the teacher (at the chalkboard) and the students (in their handbooks) do examples or activities *together*. The students follow the process by modeling the same visual activity presented in their *Algebra Ready* handbooks. This provides hands-on learning and a resource for review or future reference.

6. Guided Class Practice

Following the Guided Learning, and when the teacher feels assured that the students are ready to work independently, Guided Class Practice is introduced. The teacher walks through the class as the students work the problems, giving immediate reinforcement to students who have correct responses by marking individual problems with a **C**, ★, or ✔. This method instills confidence and encourages student productivity. This is also an opportunity to make positive verbal comments to each student and to give *individual assistance* to any student who may be having difficulty. This support and assistance to students in need of help is immediate. The guided practice provides assurance to the teacher that students are ready to move on independently with a high degree of success. Where student learning is not evident during guided practice, reteaching is required.

7. Independent Practice

This component of *Algebra Ready* immediately reinforces the day's lesson. Because students have had the opportunity to learn skills and concepts thoroughly, the Independent Practice produces a high degree of success and perpetuates student self-confidence and proof that one can be successful in math. When the *entire* instructional approach of *Algebra Ready* is used, student success rates of at least 85% are expected.

8. Try This

Periodically throughout the handbook, a Try This activity is included to provide students with a challenge and to add some algebraic concepts and problem-solving opportunities to their *Algebra Ready* experiences. Try This is used according to teacher discretion and student readiness.

Additional Information

Several additional activities and resources are built into *Algebra Ready* for teachers to use to maintain, strengthen, and expand the program. These special components include the following:

1. *Algebra Ready* bookmark
2. Detailed explanation for order of operations
3. Vocabulary words by lesson
4. Diagnostic assessment subtests
5. Sample diagnostic assessment report
6. Diagnostic assessment report forms
7. Skill drills to reinforce addition, subtraction, multiplication, and division skills
8. Forty-eight additional Word Problem Warm-Ups (24 in Part 1 and 24 in Part 2)
9. Glossary of math terms
10. Scope and sequence
11. Vocabulary card holder pattern
12. Student completion award
13. Math vocabulary cards

Algebra Ready
Lesson 1

① Word Problem Warm-Up

At Ted's Computer Shop you can buy a monitor for $895, a hard drive for $1,907, and a printer for $828. At Tom's Computer Shop the ⓢame equipment costs $3,398.

a. ⒺstimateⒺ which is the better deal. **Tom's Computer Shop** is the better deal.

b. What is the ⒺxactⒺ total price at Ted's Computer Shop? **$3,630**

c. What is the ⒹifferenceⒹ in price between the two shops? **$232**

Show work here.

```
    ④  895 ⎫                    ⑤  3,630   Ted
      1,907 ⎬ Ted's                - 3,398  Tom
    +   828 ⎭  Shop                  $232
       $3,630
```

Problem-solving plan...	
Read	1
Key words	2
Analyze	3
Add	4
Subtract	5
Multiply	
Divide	
Compare/Check	

② LESSON 1: OBJECTIVES

1. Perform the order of operations: × ÷ and + − only.
2. Practice addition, subtraction, multiplication, and division.

③ Vocabulary

1. ***sum*** — The sum is the answer to an addition problem.
2. ***difference*** — The difference is the answer to a subtraction problem.
3. ***product*** — The product is the answer to a multiplication problem.
4. ***quotient*** — The quotient is the answer to a division problem.
5. ***order of operations*** — The order of operations gives the rules for doing a problem that requires more than one operation.

Note: Italicized words appear in the lesson.

 **Lesson Presentation**

1. Perform the order of operations: × ÷ and + − only.
2. Practice addition, subtraction, multiplication, and division.

Order of Operations

Sometimes a math problem involves using more than one operation to compute the correct answer. To solve this type of problem, mathematicians have agreed upon a certain order in which to do the work. It is called the order of operations. The order of operations is like a recipe.

1. First, multiply (find the product) or divide (find the quotient) in order from left to right, just like reading.
2. Next, add (find the sum) or subtract (find the difference) in order from left to right.

The phrase **My** **D**ear **A**unt **S**ally is a device to help remember the correct order of operations.

My	**M**ultiplication	} Do these two operations, in whichever order they appear, working from left to right.
Dear	**D**ivision	
Aunt	**A**ddition	} Then do these two operations, in whichever order they appear, working from left to right.
Sally	**S**ubtraction	

Examples:

$17 - 12 \div 2 \times 2 + 8$	Divide (12 ÷ 2) first because it appears first left to right.
$17 - 6 \times 2 + 8$	Now multiply (6 × 2) before you add or subtract.
$17 - 12 + 8$	Subtract (17 − 12) because it is the first operation on the left.
$5 + 8 = 13$	Now you add to get the answer.

$7 - 5 + 2 - 3$	Subtract (7 − 5) first because it appears first left to right.
$2 + 2 - 3$	Now add (2 + 2) because it appears next.
$4 - 3 = 1$	Finally, do the last operation to get the answer.

Note: For further information, see Explanation: Order of Operations on Resources page 137.

Algebra Ready

Lesson 1

5 Guided Learning

Definition or Problem	Notes and Explanations
$4 \times 7 + 2 \times 3$	*Multiply left to right first.*
$28 + 2 \times 3$	
$28 + 6$	*Add.*
34	
$18 \div 6 + 2 \times 9 - 5$	*Divide.*
$3 + 2 \times 9 - 5$	*Multiply.*
$3 + 18 - 5$	*Add.*
$21 - 5$	*Subtract.*
16	
$3 \times 8 - 6 + 4 \times 2 + 7 - 8 \times 2$	*Multiply left to right first.*
$24 - 6 + 4 \times 2 + 7 - 8 \times 2$	
$24 - 6 + 8 + 7 - 8 \times 2$	
$24 - 6 + 8 + 7 - 16$	*Add and subtract left to right.*
$18 + 8 + 7 - 16$	
$26 + 7 - 16$	
$33 - 16$	
17	

Definition or Problem	Notes and Explanations

6 Guided Class Practice

My **D**ear **A**unt **S**ally

Let's try these examples.

1. $7 + 24 \div 4 - 3$ **10**

2. $12 \div 2 \times 3 - 18 + 4$ **4**

3. $7 + 7 \times 7$ **56**

4. $9/3 + 4 \times 4 - 2$ **17**

Note: / means divide.

Algebra Ready

Lesson 1

Name _____

Date _____

Period _____

7 Independent Practice

My **D**ear **A**unt **S**ally

Solve the following problems. Show all work and circle your answers.

1. $5 + 5 \times 5$ **30**

2. $8 + 4 + 6 - 3 - 2 - 5$ **8**

3. $4 + 4 \times 4$ **20**

4. $15 - 8 + 6 - 4 + 3 - 9$ **3**

   Hint: Subtraction is the first operation in this problem.

5. $12 \div 4 \times 6 + 8 \times 2 + 1$ **35**

6. $16 \times 2/4 \times 3/6/2$ **2**

   Note: Remember, / means divide.

7. $4 + 4 \times 4 + 4 \times 4$ **36**

8. $14 - 3 - 5 - 1 - 2 - 3$ **0**

9. $2 \times 3 \times 4/2/3/2$ **2**

10. $1 + 1 \times 1 - 1/1 \times 1 + 1$ **2**

11. $5 \times 4 \div 2 \times 4 \div 2 \times 2 \div 4$ **10**

12. $20/2 + 6 + 4/2 \times 3 + 8 - 10$ **20**

13. $3 + 3 \times 3 \times 3 + 3/3 + 2$ **33**

14. $3 \times 4/2 \times 6/3 \times 4/2$ **24**

BONUS

$18 + 60/6 + 34 - 13 \times 3 + 24/2/4 \times 2$ **29**

Algebra Ready — Lesson 2

1. Word Problem Warm-Up

Alex wanted to buy a new motorcycle and some accessories. The total cost of the bike she wanted was $5,500. Alex needed to take out a loan for her purchase, so she went to several banks to apply for loans. The first bank would give her a loan for $6,300 (Loan 1). The second bank would give her a loan for $6,750 (Loan 2).

a. Compared to Loan 1, how much more will Alex have to pay back if she takes Loan 2? **$450**

b. How much more is Loan 1 than the price of the bike? **$800**

c. Find the sum of the prices of the following accessories: a helmet at $457, a saddlebag at $263, and a leather jacket at $338. **$1,058**

Show work here.

Problem-solving plan…
Read	
Key words	
Analyze	
Add	
Subtract	
Multiply	
Divide	
Compare/Check	

2. LESSON 2: OBJECTIVES

1. Perform the order of operations: exponents, × ÷, and + −.
2. Practice addition, subtraction, multiplication, and division.

3. Vocabulary

1. *exponent* — The exponent is the small raised number to the upper right of the base number that tells you how many times to multiply the base by itself.

2. *base* — The base, in terms of exponents, is the number in front of the exponent. The base is multiplied by itself as many times as the exponent says.

3. *power* — Power is another name for exponent.

4. *estimate* (verb) — To estimate is to make an educated guess based on information in a problem.

Note: Remember, italicized words appear in the lesson.

Lesson Presentation

1. Perform the order of operations: exponents, × ÷, and + −.
2. Practice addition, subtraction, multiplication, and division.

Order of Operations

Exponents (powers) are now added to the order of operations.

1. First, solve all exponents.
2. Next, multiply or divide from left to right.
3. Last, add or subtract from left to right.

Remember **M**y **D**ear **A**unt **S**ally? Now let's add the word **E**xcuse so we can remember exponents also.

Excuse	**E**xponents	Solve exponents first.
My	**M**ultiplication	} Next, × and ÷ are performed in whichever order they
Dear	**D**ivision	appear.
Aunt	**A**ddition	} Then + and − are performed in whichever order they
Sally	**S**ubtraction	appear.

Example:

$6^2 + 12 \div 2 \times 3 - 8 + 4$ First, solve the exponent or power (6^2).

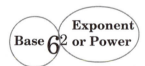

The base tells *what number* to multiply by itself.
The exponent tells *how many times* to multiply the base by itself.

$36 + 12 \div 2 \times 3 - 8 + 4$ Next, divide 12 by 2 because, reading left to right, it appears first.

$36 + 6 \times 3 - 8 + 4$ Multiply 6 by 3.

$36 + 18 - 8 + 4$ Add and subtract from left to right.

$54 - 8 + 4$

$46 + 4 = 50$

Algebra Ready — Lesson 2

5 Guided Learning

Definition or Problem	Notes and Explanations
	Note: Students may need to review working with exponents.
$3^2 + 6 \div 2 + 18 - 5$	**Exponents are worked first.**
$9 + 6 \div 2 + 18 - 5$	**Divide.**
$9 + 3 + 18 - 5$	**Add and subtract left to right.**
$12 + 18 - 5$	
$30 - 5$	
25	
$3 + 6 \times 4^2 + 5 - 3$	**Exponents are worked first.**
$3 + 6 \times 16 + 5 - 3$	**Multiply.**
$3 + 96 + 5 - 3$	**Add and subtract left to right.**
$99 + 5 - 3$	
$104 - 3$	
101	
$75 \div 25 \times 4 - 5$	**There are no exponents. In this problem, division is the first multiplication or division operation.**
$3 \times 4 - 5$	**Now multiply.**
$12 - 5$	**Subtract.**
7	

Definition or Problem	Notes and Explanations

6 Guided Class Practice

Excuse **M**y **D**ear **A**unt **S**ally

Let's try these examples.

1. $3^2 \times 5 + 15/5$ **48**

   Note: Remember, / means divide.

2. $8 \times 4 - 4^2 + 14/2$ **23**

3. $16 \div 4 + 4^3 \times 5 - 7$ **317**

4. $18 \div 3 \times 5 + 4 - 2 + 6^3$ **248**

Algebra Ready
Lesson 2

Name _____
Date _____
Period _____

 Independent Practice

Excuse **M**y **D**ear **A**unt **S**ally

Solve the following problems. Show all work and circle your answers.

1. $6 \div 2 + 3 \times 3$ *12*

2. $4^2 - 5 + 3^2 + 2^4$ *36*

3. $1 + 1 \times 1 + 1 \times 1 + 1 \times 1$ *4*

4. $6^2 - 5^2 + 4^2 - 3^2 + 2^2$ *22*

5. $3 \times 6/9 + 8 \times 4/2$ *18*
   Note: Remember, / means divide.

6. $2^2 \times 2^3 \times 2^4$ *512*

7. $2^4 + 2^3$ *24*

8. $3^3 + 4^3$ *91*

9. $3^3 \times 3^2$ **243**

10. $2 + 2^3 + 3^4 - 5^2$ **66**

11. $3^3 + 3^3/3^2$ **30**

12. $2^3 \times 3^2/6^2 + 4/2^2$ **3**

13. $8 \div 2 + 4^2 \times 4 + 4^3/4$ **84**

14. $2^3 + 3^3 + 4^3 - 9^2$ **18**

15. $2^2 + 2^3 + 2^4$ **28**

16. $3^5 + 3^4 \times 3^3$ **2,430**

17. $3^2 + 2^2 \times 6/2 - 3^2$ **12**

18. $2^3 + 3^2 \times 2^4 - 81$ **71**

BONUS

Place the four math operators (×, ÷, +, and −) in the blank spaces to make the equation true. You will use each operator once.

$$3 \underline{\;+\;} 6 \underline{\;\div\;} 2 \underline{\;\times\;} 4 \underline{\;-\;} 6 = 9$$

Algebra Ready — Lesson 3

1. Word Problem Warm-Up

Each student has four textbooks, two notebooks, three pencils, and six sheets of paper. There are 24 students.

a. How many textbooks are there? **96**

b. How many notebooks are there? **48**

c. How many pencils are there? **72**

d. How many sheets of paper are there? **144**

Show work here.

Problem-solving plan…
- Read
- Key words
- Analyze
- Add
- Subtract
- Multiply
- Divide
- Compare/Check

2. LESSON 3: OBJECTIVE

Perform the order of operations: parentheses and other grouping symbols (brackets and braces), exponents, × ÷, and + −.

3. Vocabulary

1. **squared** — Squared refers to an exponent (power) of two (^{2}).

2. **cubed** — Cubed refers to an exponent (power) of three (^{3}).

3. **percent** — Percent is a way to represent part of a whole that has been divided into 100 equal pieces.

4. *expression* — An expression is a mathematical statement that stands for a given value.

5. *grouping symbols* — Grouping symbols are symbols such as parentheses ( ), brackets [ ], or braces { } that show you what to do first in a mathematical expression.

 Lesson Presentation

> Perform the order of operations: parentheses and other grouping symbols (brackets and braces), exponents, × ÷, and + −.

Order of Operations

Parentheses and other grouping symbols now become a part of the order of operations. To find the value of an expression with grouping symbols, follow these steps.

1. Perform the order of operations inside the parentheses or other grouping symbols first. When grouping symbols are nested { [( )] }, work from the inside to the outside.
2. Solve the exponent(s).
3. Multiply or divide from left to right.
4. Finally, add or subtract from left to right.

Parentheses are the most common grouping symbols. To remember parentheses and other grouping symbols, let's add the word **P**lease to our phrase **E**xcuse **M**y **D**ear **A**unt **S**ally.

Please	**P**arentheses	Parentheses and other grouping symbols are done first.
Excuse	**E**xponents	Exponents are solved next.
My **D**ear	**M**ultiplication **D**ivision	} Done in the order they appear.
Aunt **S**ally	**A**ddition **S**ubtraction	} Done in the order they appear.

Example:

$(12 - 7)^2 \div 5 \times (19 + 1) - 1$	Subtract within the left set of parentheses first.
$5^2 \div 5 \times (19 + 1) - 1$	Add within the remaining set of parentheses.
$5^2 \div 5 \times 20 - 1$	Solve the exponent.
$25 \div 5 \times 20 - 1$	Multiply and divide left to right. Here, division is first.
$5 \times 20 - 1$	Then multiply.
$100 - 1 = 99$	Last, do the subtraction.

Algebra Ready

Lesson 3

5. Guided Learning

Definition or Problem	Notes and Explanations
$(4 - 3)^2 + (6 \div 2 + 4) \times 7$	Work inside the left set of parentheses first.
$1^2 + (6 \div 2 + 4) \times 7$	Divide, then add, inside the remaining set of parentheses.
$1^2 + (3 + 4) \times 7$	
$1^2 + 7 \times 7$	Solve the exponent.
$1 + 7 \times 7$	Multiply.
$1 + 49$	Add.
50	
	Notice the nested grouping symbols.
$6 \div 3 + 2[(3 + 7) \div 2]$	Work inside the parentheses first.
$6 \div 3 + 2[10 \div 2]$	Now work inside the brackets.
$6 \div 3 + 2 \times 5$	Divide, then multiply.
$2 + 10$	Do the addition.
12	

Definition or Problem	Notes and Explanations

⑥ Guided Class Practice

Please Excuse My Dear Aunt Sally

Let's try these examples.

1. $8 + 24 \div (4 - 3)$ **32**

2. $[(6 \times 3) \div 3^2 - 1] + 2 \times 4$ **9**

3. $4(2 + 2^2)$ **24**

   Note: A number right next to a grouping symbol means to multiply.

4. $2[18 \div (4 + 2) + 1^3]$ **8**

Algebra Ready

Lesson 3

Name _____

Date _____

Period _____

7 Independent Practice

Please **E**xcuse **M**y **D**ear **A**unt **S**ally

Find the value of the following expressions. Show all work and circle your answers.

1. $3 + 3 \times 3$ *12*

2. $6(2+3)^2$ *150*
   Note: A number right next to a grouping symbol means to multiply.

3. $64 \times 2/4/2/2/4$ *2*
   Note: Remember, / means divide.

4. $3(2+3^2)$ *33*

5. $6 \div 2(2+1)$ *9*

6. $(11-3)^2$ *64*

7. $(3+2)(6-3)$ *15*

8. $(8-4)^2$ *16*

9. $7(13 - 4) + 7$ **70**

10. $6 \div 2 + 3(8 - 3) + 3^2$ **27**

11. $6 \div 2(8 + 7 - 10)^2$ **75**

12. $4(2 + 3) + 5(6 + 4)$ **70**

13. $5(3 + 4)^2$ **245**

14. $3(3 + 3 \times 3)$ **36**

15. $17(2^3 - 8)$ **0**

16. $4[3 + 2(3 + 4) - 11]$ **24**

17. $2(11 - 3^2 + 2)^2$ **32**

18. $12 \div 2[4 + 3(3 + 2)]$ **114**

BONUS

$18 + (60 \div 10) - 2 + 125/5^3 - (16 - 3^2)$ **16**

Algebra Ready — Lesson 4

1. Word Problem Warm-Up

A skateboard is on sale for $72.00. If the sales tax is 7% (7 percent), what is the total cost of the skateboard? __$77.04__

Show work here.

Problem-solving plan...
- Read
- Key words
- Analyze
- Add
- Subtract
- Multiply
- Divide
- Compare/Check

2. LESSON 4: OBJECTIVE

Review the order of operations: parentheses and other grouping symbols (brackets and braces), exponents, × ÷, and + −.

Please **E**xcuse **M**y **D**ear **A**unt **S**ally

3. Vocabulary

1. **simplify** — To simplify is to combine like terms and put an answer in its lowest form.

2. **order of operations** — The order of operations gives you the rules for doing a problem that requires more than one operation.

3. **variable** — A variable is a symbol used to take the place of an unknown number.

4. **substitute** (verb) — To substitute means to replace a variable or symbol with a known numerical value.

5. **evaluate** — To evaluate is to find the value of an expression once the values of the variables are known.

4 Lesson Presentation

> Review the order of operations: parentheses and other grouping symbols (brackets and braces), exponents, × ÷, and + −.
>
> **Please Excuse My Dear Aunt Sally**

Order of Operations

To find the value of an expression, follow the order of operations.

1. Perform the operations inside the parentheses or other grouping symbols first. When grouping symbols are nested, work from the inside to the outside.
2. Solve the exponent(s).
3. Multiply or divide from left to right.
4. Finally, add or subtract from left to right.

Example:

$[2 + (4 − 1) − 1] \times 3 + 2^2 \times 2$	First work from the inside out in the nested grouping symbols. Subtract within the parentheses.
$[2 + 3 − 1] \times 3 + 2^2 \times 2$	Now work within the brackets.
$[5 − 1] \times 3 + 2^2 \times 2$	
$4 \times 3 + 2^2 \times 2$	Solve the exponent.
$4 \times 3 + 4 \times 2$	Next, multiply left to right.
$12 + 4 \times 2$	
$12 + 8 = 20$	Add.

5 Guided Learning

Definition or Problem	Notes and Explanations

Algebra Ready — Lesson 4

Definition or Problem	Notes and Explanations
	Work from the inside out. There is nothing to do inside the parentheses, so solve the exponent that is next to
$[3 + 2(2)^2]^2$	the parentheses.
$[3 + 2(4)]^2$	Finish working inside the brackets. Multiply first.
$[3 + 8]^2$	Add.
11^2	Solve the exponent.
121	
$25 \div 5^2 + (2 + 1)^2 - 4$	Work inside the parentheses first.
$25 \div 5^2 + 3^2 - 4$	Solve the exponents.
$25 \div 25 + 3^2 - 4$	
$25 \div 25 + 9 - 4$	Divide.
$1 + 9 - 4$	Add.
$10 - 4$	Subtract.
6	
$6^2 - 2^4 - 4^2 + 3^3$	Work each exponent.
$36 - 16 - 16 + 27$	Subtract and add left to right.
$20 - 16 + 27$	
$4 + 27$	
31	

Definition or Problem	Notes and Explanations

6 Guided Class Practice

Please Excuse My Dear Aunt Sally

Let's try these examples.

1. $(8 + 24) \div 4^2 - 2$ **0**

2. $7 + (7^2 \div 7) + 7$ **21**

3. $(6 + 4) \div (3^2 + 1)$ **1**

4. $18 \div (4 + 2) + 1^3$ **4**

Algebra Ready

Lesson 4

Name _____

Date _____

Period _____

⟨7⟩ Independent Practice

Please **E**xcuse **M**y **D**ear **A**unt **S**ally

Find the value of the following expressions. Show all work and circle your answers.

1. $3 + 3 \times 3$ **12**

2. $64/2/4/8 \times 5$ **5**

3. $12 \div 2 \times 4 + 6 \times 3/9$ **26**

4. $3^4 - 5$ **76**

5. $3(3 + 2)^2$ **75**

6. $5(3 - 1)^2$ **20**

7. $6(9 + 5 - 10)^2$ **96**

8. $(8 - 5)(11 + 7)$ **54**

9. $4(8-4)^3$ *256*   **10.** $(3+2^3)(2+3^2)$ *121*

11. $4[(4-2)+4^2]-15$ *57*   **12.** $7(4+5)^2$ *567*

13. $6(3+3)^2-3(6+3)$ *189*   **14.** $1+1(1+1)+1\times 1/1$ *4*

15. $2+2(2+2)^3$ *130*   **16.** $5+3(8-5)^3+6$ *92*

17. $6+3(2+1)^4-50$ *199*   **18.** $1+1[1+1(1+1)]+1$ *5*

19. 3^3-3^2-3 *15*   **20.** $3^4-2^5-5^2$ *24*

Algebra Ready

Lesson 4

Name _____
Date _____
Period _____

7 More Independent Practice

Please **E**xcuse **M**y **D**ear **A**unt **S**ally

Find the value of the following expressions. Show all work and circle your answers.

1. $(4^2 + 3)^2$ *361*

2. $7(5^2 - 4^2)$ *63*

3. $[2 + 2(2)^2]^2$ *100*

4. $5 + 3(2)^3 - 5$ *24*

5. $4(2 + 5)^2 - 60$ *136*

6. $6[2 + 2(2)]$ *36*

7. $6(2 + 2^2)^2$ *216*

8. $2 + [2 + 2(2)^2]^2$ *102*

9. $2 + 2(3)^4 - 3(5)^2$ **89**

10. $9(3 + 4^2 - 9)^2$ **900**

11. $6[2 + 2(3)^2 - 16] - 10$ **14**

12. $2^4 + 2^2 + 2^3$ **28**

13. $2 + 2^2[2 + 2(2 + 2) + 2] + 2^2$ **54**

14. $3^2(2^3 - 2^2 + 5)^2$ **729**

15. 5^6 **15,625**

16. 7^4 **2,401**

17. $2^3(2^2 + 3^3) - 3^5$ **5**

18. $2[2 + 2(2 + 2)^2]^2$ **2,312**

BONUS

$7^3 - 6^3 + (5^3 - 4^3) + (3^3 - 2^3 + 1^3)$ **208**

Algebra Ready

Try This 8

Symbols Into Words

Express the following in words. *Answers may vary.*
Note: Numbers do not have to be written as words.

1. $2 + 4$

   2 plus 4; the sum of 2 and 4; 2 added to 4

2. $6 - 3$

   6 minus 3; the difference of 6 and 3; 3 less than 6

3. 4×2

   4 times 2; the product of 4 and 2; 4 doubled

4. $8 \div 2$

   8 divided by 2; the quotient of 8 and 2

5. $6 - y$

   the difference between 6 and y; 6 minus y

6. $\dfrac{b}{3}$

   b over 3; b divided by 3

7. $3x$

   3 times x; the product of 3 and x

8. x^2

   x squared; x to the second power

Express the following in words.

9. $5x + 4$

the product of 5 and x added to 4

10. $3 \cdot 4x$

3 times the product of 4 and x

11. $4(n + 1)$

4 times the sum of n and 1

12. $3x^2$

x squared times 3

13. $\dfrac{n + 4}{3}$

n plus 4 all over 3; the sum of n and 4 divided by 3

14. $x + y - z$

x plus y minus z; x increased by y then decreased by z

15. $\dfrac{x}{x - 3}$

x divided by the difference of x and 3

16. $3x + 2 = 7$

the sum of 3 times x and 2 equals 7

17. $a + b^2 + c^3 = x$

the sum of a, b squared, and c cubed equals x

Algebra Ready — Lesson 5

1. Word Problem Warm-Up

Cathy bought school supplies at Suzanne's School Supply Shop. Her supplies totaled $35.92. If the sales tax is 4.5%, what is the total cost for her school supplies? **$37.54**

Show work here.

Problem-solving plan...
- Read
- Key words
- Analyze
- Add
- Subtract
- Multiply
- Divide
- Compare/Check

2. LESSON 5: OBJECTIVE

Evaluate formulas, specifically the area of a circle. $A = \pi r^2$

3. Vocabulary

1. **equation** — An equation is a mathematical sentence that has two equivalent, or equal, expressions separated by an equal sign (=).

2. **formula** — A formula is a recipe or equation used to find specific information.

3. **area** — The area is the space inside a closed, two-dimensional figure.

4. **diameter** — The diameter is a line segment across a circle that goes through the center and divides the circle into two equal halves.

5. **radius** — The radius is a line segment from the center point of a circle to any point on the circle. It is half the length of the diameter.

6. **circumference** — The circumference is the boundary line of a circle.

7. **pi (or π)** — Pi is approximately equal to $\frac{22}{7}$, or 3.14. It is the ratio of the circumference of any circle to the diameter of the circle.

4 Lesson Presentation

> Evaluate formulas, specifically the area of a circle. $A = \pi r^2$

Evaluate the Formula for the Area of a Circle

To evaluate (solve) the formula for the area of a circle, follow these steps.

1. Substitute the given values for the symbols in the formula.
2. Follow the order of operations to evaluate the formula.
3. Label your final answer.

Example:

Find the area (A) of a circle with a radius (r) of 4 cm. Use 3.14 for π.

$\boxed{A = \pi r^2}$	This is the formula (equation) for the area of a circle.
$A = \pi r^2$	Substitute the values from the diagram into the formula. Follow the order of operations to evaluate the formula (solve the equation).
$A = 3.14(4^2)$	Solve the exponent.
$A = 3.14(16)$	Multiply.
$A = 50.24 \text{ cm}^2$	The area is labeled in square centimeters because cm · cm = cm².

r = 4 cm

5 Guided Learning

Definition or Problem	Notes and Explanations
$A = \pi r^2$	r = 2.4 cm
$A = \pi r^2$	Substitute 3.14 for π and 2.4 for r.
$A = 3.14(2.4^2)$	Follow the order of operations. Solve the exponent first.
$A = 3.14(5.76)$	Multiply.
$A = 18.086 \text{ cm}^2$	Explain why cm² is the proper label. cm · cm = cm²

Algebra Ready
Lesson 5

Definition or Problem	Notes and Explanations
$A = \pi r^2$	$r = 80$ in
$A = \pi r^2$	Substitute 3.14 for π and 80 for r.
$A = 3.14(80^2)$	Follow the order of operations. Solve the exponent first.
$A = 3.14(6,400)$	Multiply.
$A = 20,096$ in^2	Label the answer.

Definition or Problem	Notes and Explanations

6 Guided Class Practice

Let's try these examples. Use the formula $A = \pi r^2$ to find the area (A) of circles with the following radii. Use 3.14 as the value of π.

1. r = 2 in   *12.56 in²*

2. r = 3.5 mm   *38.465 mm²*

3. r = 1 cm   *3.14 cm²*

4. r = 1,000 in   *3,140,000 in²*

Algebra Ready   Lesson 5

Name _____

Date _____

Period _____

⟨7⟩ Independent Practice

Find the area (A) of circles with the following radii. Use the formula $A = \pi r^2$, and use 3.14 as the value of π. Show all work and circle your answers.

1. r = 5 mm   **78.5 mm²**   2. r = 8 ft   **200.96 ft²**

3. r = 9 cm   **254.34 cm²**   4. r = 15 cm   **706.5 cm²**

5. r = 16 in   **803.84 in²**   6. r = 20 ft   **1,256 ft²**

7. r = 25 m   **1,962.5 m²**   8. r = 30 mm   **2,826 mm²**

9. r = 50 ft   **7,850 ft²**   10. r = 100 cm   **31,400 cm²**

11. r = 1.1 in   *3.799 in²*   12. r = 1.2 m   *4.522 m²*

13. r = 2.5 mm   *19.625 mm²*   14. r = 10 in   *314 in²*

15. r = 8.8 m   *243.162 m²*   16. r = 9.6 cm   *289.382 cm²*

17. r = 21.3 cm   *1,424.587 cm²*   18. r = 3.1 cm   *30.175 cm²*

19. r = 5,000 in *78,500,000 in²*   20. r = 10,000 mm *314,000,000 mm²*

21. r = 3.141 m   *30.979 m²*   22. r = 1.11 mm   *3.869 mm²*

BONUS

Find the area of a circle with a radius of 0.75 cm.   *1.766 cm²*

Algebra Ready — Lesson 6

1. Word Problem Warm-Up

A submarine is 345 feet under the ocean surface and rises 59 feet.

a. How deep is the submarine now?   ⁻286 ft

b. How many more feet must it rise to be 50 feet below the surface?   236 ft

Show work here.

Problem-solving plan…
- Read
- Key words
- Analyze
- Add
- Subtract
- Multiply
- Divide
- Compare/Check

2. LESSON 6: OBJECTIVES

1. Combine integers using the number line.
2. Apply the first three signed number rules.

3. Vocabulary

1. **whole numbers** — Whole numbers are the counting numbers and 0 (0, 1, 2…).

2. ***integers*** — Integers are the set of counting numbers, their opposites, and 0 (…⁻2, ⁻1, 0, ⁺1, ⁺2…).

3. **opposites** — Opposites are the positive and negative of the same number.

4. ***zero (0)*** — Zero separates the positive numbers and negative numbers on a number line.

5. **number line** — A number line is a line with equal distances marked off to represent numbers.

 **Lesson Presentation**

> 1. Combine integers using the number line.
> 2. Apply the first three signed number rules.

Number Line

Three Rules for Combining Signed Numbers

Follow these three rules to combine positive and negative numbers.

Combination of Numbers	Operation	Answer Looks Like
1. Positive, Positive	+	Positive
2. Negative, Negative	+ (Note: The "+" sign doesn't actually show up! Instead of ⁻5 + (⁻6), the problem will be written ⁻5 – 6. You add the two negative numbers and get ⁻11.)	Negative
3. Positive, Negative (or Negative, Positive)	– (Note: Always subtract when combining positive, negative or negative, positive numbers.)	The sign of the "larger" number (The number with the highest absolute value)

Examples:

1. $4 + 5 = 9$ Rule 1.
2. $⁻5 – 6 = ⁻11$ Rule 2.
3. $18 – 22 = ⁻4$ Rule 3.

    $⁻8 + 3 + 5$ Rule 3. Add or subtract left to right.
       $⁻5 + 5 = 0$ Zero is neither positive nor negative.

Algebra Ready
Lesson 6

5. Guided Learning

Definition or Problem	Notes and Explanations
	Note: Teachers are encouraged to teach combining of signed numbers by any method that is appropriate for their specific classes.
	Combine using the signed number rules.
$-5 + 6$	**Combine using Rule 3.**
1	
	Check your answer using a number line.
	Start on -5.
	Move right ☞ six spaces. Be sure to count the 0.
	Use the rules for combining signed numbers. *Note: Add or subtract from left to right.*
$3 - 4 + 5$	**Subtract.**
$-1 + 5$	**Combine.**
4	
	Check your answer using a number line.
	Start on 3.
	Move left ☜ four spaces.
	Move right ☞ five spaces.
	Give the rules that apply to the following.
$-5 - 5$	**Negative, Negative = Negative after the numbers**
-10	**are added. (Rule 2)**
$6 - 15$	**Positive, Negative = The sign of the larger number after**
-9	**the numbers are subtracted. (Rule 3)**

Definition or Problem	Notes and Explanations
2 − 4 − 5 + 3	Add or subtract left to right following the rules for combining signed numbers.
⁻2 − 5 + 3	
⁻7 + 3	
⁻4	

Guided Class Practice

Let's try these examples. Add numbers to the number lines and use them to solve the problems.

1. ⁻4 + 9
   5
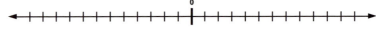

2. 5 − 10
   ⁻5
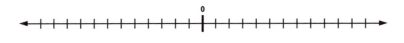

Use the rules for combining signed numbers to solve the following. Give the number of the rule(s) you use.

3. ⁻7 − 8   **⁻15 Rule 2**

4. 4 − 15   **⁻11 Rule 3**

5. ⁻7 + 10   **3 Rule 3**

6. ⁻4 + 5 + 2 − 5   **⁻2 Rule 3**
                          Rule 1
                          Rule 3

Algebra Ready — Lesson 6

Name _____

Date _____

Period _____

7 Independent Practice

Add numbers to the number line and use it to answer problems 1–4. Circle your answers. Give the number of the rule(s) that apply.

1. $^-6 + 8$ **2 Rule 3**

2. $^-3 - 5$ **$^-8$ Rule 2**

3. $4 - 7 + 6$ **3 Rule 3**

4. $^-8 - 2$ **$^-10$ Rule 2**

Use the rules for combining signed numbers to solve the following. Circle your answers.

5. $^-9 + 5$ **$^-4$**

6. $^-13 + 9$ **$^-4$**

7. $2 - 9$ **$^-7$**

8. $13 - 7$ **6**

9. $^-11 - 12$ **$^-23$**

10. $6 - 20$ **$^-14$**

11. ⁻6 − 6   *-12*   12. ⁻2 + 7 − 6   *-1*

13. ⁻4 − 3 − 2   *-9*   14. 8 − 17 − 9 + 11   *-7*

15. ⁻9 − 8 − 7   *-24*   16. ⁻6 + 5 + 10   *9*

17. 28 − 9 − 12 − 5   *2*   18. ⁻6 − 14   *-20*

19. 4 − 12   *-8*   20. 7 − 5 − 12   *-10*

21. ⁻8 + 7 − 6 + 5 − 4   *-6*   22. ⁻10 + 27 − 17   *0*

23. ⁻87 + 64   *-23*   24. ⁻14 − 9 + 6   *-17*

Algebra Ready — Lesson 7

1. Word Problem Warm-Up

Volume (V) is found by multiplying the length (l) times the width (w) times the height (h). Find the volume of the aquarium below.

$V = lwh$

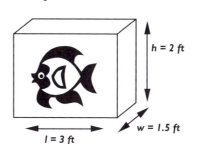

The volume is ___*9 ft³*___ .

Show work here.

Problem-solving plan...	
Read	
Key words	
Analyze	
Add	
Subtract	
Multiply	
Divide	
Compare/Check	

2. LESSON 7: OBJECTIVES

1. Apply signed number rules 1 through 6.
2. Review signed number rules 1 through 3.

3. Vocabulary

1. ***volume*** — Volume is the amount of space inside a three-dimensional object.
2. **base** — The base, in terms of geometry, generally refers to the side on which a figure rests.
3. **centimeter** — A centimeter is a small unit of metric length that is 1/100th of a meter.
4. **plane** — A plane is a flat surface, with no thickness, that extends forever.

4. Lesson Presentation

1. Apply signed number rules 1 through 6.
2. Review signed number rules 1 through 3.

Six Signed Number Rules

Here are three more rules to use when combining positive and negative numbers.

Combination of Numbers	Operation	Answer Looks Like
1. Positive, Positive	+	Positive
2. Negative, Negative	+	Negative
3. Positive, Negative (or Negative, Positive)	−	The sign of the "larger" number (The number with the highest absolute value)
4. Positive, Positive (× or ÷)	× or ÷	Positive
5. Negative, Negative (× or ÷)	× or ÷	Positive
6. Positive, Negative (× or ÷) (or Negative, Positive)	× or ÷	Negative

Note:

Parentheses have three jobs:

1. To *indicate* multiplication. $(3)(4) = 12$
2. To *clarify* the sign of a number. $(^-7) + (^-3) = {^-10}$
3. To *act* as grouping symbols. $(3 + 2)^2 + 4 =$
$5^2 + 4 =$
$25 + 4 = 29$

Example:

$(^-3)(5)(^-2)(2)$ — Multiply left to right following the signed number rules. $(-3)(5) = -15$. Rule 6

$^-15(^-2)(2)$ — $^-15(^-2) = 30$. Rule 5

$30(2)$ — $30(2) = 60$. Rule 4

60

Algebra Ready — Lesson 7

5 Guided Learning

Definition or Problem	Notes and Explanations
$10 - 5 + 4 + 2 - 1$	Add or subtract left to right following the signed number rules.
$5 + 4 + 2 - 1$	
$9 + 2 - 1$	
$11 - 1$	
10	
$(^-3)(^-2)(^-4)(^-2)$	Multiply left to right.
$6(^-4)(^-2)$	
$^-24(^-2)$	
48	
$(^-3)^2 + (2 - 7)$	Work inside the parentheses that are grouping symbols first.
$(^-3)^2 + (^-5)$	Solve the exponent.
$9 + (^-5)$	Combine the signed numbers using Rule 3.
4	

Definition or Problem	Notes and Explanations

⑥ Guided Class Practice

Let's try these examples.

1. (8)(9)   **72**

2. 24 ÷ (⁻8)   **⁻3**

3. (8)(⁻1)(⁻3)(⁻1)(⁻1)   **24**

4. ⁻2 + 4 × 15   **58**

Algebra Ready
Lesson 7

Name _____

Date _____

Period _____

7 Independent Practice

Simplify the following problems. Show all work and circle your answers.

1. $^-6 - 4 - 3 - 5 - 7 - 8 - 6$ **-39**

2. $7 + 12 + 8 + 10 + 5 + 7$ **49**

3. $^-13 - 15 - 14 - 11 - 16$ **-69**

4. $20 - 15 + 17 - 6 - 13 - 7$ **-4**

5. $111 - 11 + 101 - 1,011 + 10$ **-800**

6. $47 + 36 - 28 + 30 - 27 - 40$ **18**

7. $(^-5)(^-8)$ **40**

8. $(^-4)(^-7)$ **28**

9. (12)(⁻10)   **⁻120**

10. (⁻2)(⁻5)(2)(4)   **80**

11. (⁻5)(⁻5)(⁻5)(2)   **⁻250**

12. (⁻2)(⁻3)(⁻2)(⁻4)(⁻2)(⁻3)   **288**

13. (⁻32) ÷ (⁻4)   **8**

14. ⁻36 ÷ (⁻9)   **4**

15. 42 ÷ 6   **7**

16. (⁻12)(⁻15)   **180**

17. (⁻1)(⁻1)(⁻2)(⁻3)(⁻1)(⁻1)(⁻4)   **⁻24**

18. ⁻8 − 6 + 7 + 9 + 13 − 4 + 16   **27**

19. (⁻2)(⁻4)(⁻5)(6)(3)(2)(⁻1)   **1,440**

20. ⁻100 ÷ (⁻20)   **5**

Algebra Ready — Lesson 8

1. Word Problem Warm-Up

a. On the lines below, write the following integers in sequence from least to greatest.

b. Circle the integers that have a 7 in the tens place.

c. Underline the integers that have a 7 in the hundreds place.

3,875 -2,273 4,878 -7,765 -3,849 5,749 -6,948 -4,483 -1,094 0

<u>-7,765</u>   -6,948   -4,483   -3,849   (-2,273)   -1,094

0   (3,875)   (4,878)   <u>5,749</u>

Show work here.

Problem-solving plan...
Read	
Key words	
Analyze	
Add	
Subtract	
Multiply	
Divide	
Compare/Check	

2. LESSON 8: OBJECTIVES

1. Apply the six signed number rules.
2. Review the order of operations.

3. Vocabulary

1. **sequence** — The sequence of a set of items is the order in which the items are arranged.

2. **negative number** — A negative number is any number found to the left of zero (0) on a number line.

3. **digit** — A digit is any single whole number 0 through 9.

4. **hundreds place** — The hundreds place is third from the last digit in a whole number (4,<u>6</u>51).

5. **tens place** — The tens place is second from the last digit in a whole number (4,6<u>5</u>1).

Lesson Presentation

> 1. Apply the six signed number rules.
> 2. Review the order of operations.

Six Signed Number Rules

Combination of Numbers	Operation	Answer Looks Like
1. Positive, Positive	+	Positive
2. Negative, Negative	+	Negative
3. Positive, Negative (or Negative, Positive)	−	The sign of the "larger" number (The number with the highest absolute value)
4. Positive, Positive (× or ÷)	× or ÷	Positive
5. Negative, Negative (× or ÷)	× or ÷	Positive
6. Positive, Negative (× or ÷) (or Negative, Positive)	× or ÷	Negative

> *Remember!*
>
> Parentheses have three jobs. Give an example of each.
>
> 1. To *indicate* multiplication. _Answers will vary._
>
> 2. To *clarify* the sign of a number. _Answers will vary._
>
> 3. To *act* as grouping symbols. _Answers will vary._

Example:

Use the order of operations and the six rules for combining signed numbers to solve the problem.

$8 - 4(2 - 9)^2 - 11$ Work inside the parentheses first.

$8 - 4(^-7)^2 - 11$ Solve the exponent.

$8 - 4(49) - 11$ Multiply.

$8 - 196 - 11$ Subtract from left to right.

$^-188 - 11 = {}^-199$

Algebra Ready — Lesson 8

5. Guided Learning

Definition or Problem	Notes and Explanations
	Follow the order of operations.
$7(2-4) + 3^2$	*Work inside the parentheses first.*
$7(^-2) + 3^2$	*Solve the exponent next.*
$7(^-2) + 9$	*Multiply.*
$^-14 + 9$	*Add.*
$^-5$	
$(^-2)^3 + (7-11)^2 + 10$	*Work inside the parentheses first.*
$(^-2)^3 + (^-4)^2 + 10$	*Next, solve the exponents.*
$^-8 + 16 + 10$	*Add left to right.*
$8 + 10$	
18	

Definition or Problem	Notes and Explanations

6. Guided Class Practice

Let's try these examples.

1. $(^-4)(5) + (^-6)(^-7)$ **22**

2. $^-2(4 \times 15)$ **-120**

3. $8 + (^-2)(^-3)(1)(^-2)$ **-4**

4. $10 + 3(4 - 8)^2$ **58**

Algebra Ready — Lesson 8

Name _____

Date _____

Period _____

7 Independent Practice

Simplify the following problems. Show all work and circle your answers.

1. $(-6)(-4) + (-5)(-7)$ **59**

2. $(-3)^3$ **-27**

3. $-45 + (2)(3)$ **-39**

4. $-10(4 - 8)^2$ **-160**

5. $-3 - 4 - 5 - 6 - 7 + 18$ **-7**

6. $-8 + 3(-4 + 8) - 1$ **3**

7. $-3 - 2(-1 - 3)^2$ **-35**

8. $(-42) \div (-7)$ **6**

9. $-2(-10 + 7)^2$ *-18*   **10.** $3 - 3(5 - 8)$ *12*

11. $-64 \div 8$ *-8*   **12.** $(-4)(-3)(-2)(-1)(2)(-3)$ *-144*

13. $(-3)(-4) + (-6)(2)$ *0*   **14.** $-4(-2)^4$ *-64*

15. $-2 - 4(-5 + 2)^2 - 6$ *-44*   **16.** $(-2)(-5) + 6(-4)$ *-14*

17. $(-1)(-1)(-2)(-3)(-1)(-1)(-4)$ *-24*   **18.** $7 - 6 - 4(-5 + 9) + 12 - 8$ *-11*

19. $(-4)^2$ *16*   **20.** $-14 - 16 - 15 - 12 - 13 - 11$ *-81*

Algebra Ready — Lesson 9

1. Word Problem Warm-Up

The temperature in Antarctica can fall to ⁻120° Fahrenheit (F). The temperature in Arizona has been known to reach 113° F. What is the temperature range between these two places? ____233° F____

The highest temperature ever recorded in the weather data for a particular city was 105° F. The lowest temperature recorded for that city was ⁻36° F. Find the range between these two temperatures.

____141° F____

Show work here.

Problem-solving plan...
Read	
Key words	
Analyze	
Add	
Subtract	
Multiply	
Divide	
Compare/Check	

2. LESSON 9: OBJECTIVES

1. Apply the six signed number rules including eliminating a double sign.
2. Review the order of operations.

3. Vocabulary

1. ***range*** — The range is the difference between the greatest and least values in a set of data.
2. ***data*** — Data are numerical information that can be put into charts, tables, etc., for comparison.
3. ***addend*** — An addend is a number being added to another.
4. ***equivalent*** — Equivalent refers to terms or expressions that have equal values.
5. ***order of operations*** — The order of operations gives you the rules for doing a problem that requires more than one operation.

 **Lesson Presentation**

> 1. Apply the six signed number rules including eliminating a double sign.
> 2. Review the order of operations.

Six Signed Number Rules and Eliminating a Double Sign

Sometimes a double sign (for example, + (⁻5) or − (⁻4)) will appear in a problem. To eliminate the double sign, use signed number rules 4–6 and follow the steps that are given in the box below.

Combination of Numbers	Operation	Answer Looks Like
1. Positive, Positive	+	Positive
2. Negative, Negative	+	Negative
3. Positive, Negative (or Negative, Positive)	−	The sign of the "larger" number (The number with the highest absolute value)
4. Positive, Positive (× or ÷)	× or ÷	Positive
5. Negative, Negative (× or ÷)	× or ÷	Positive
6. Positive, Negative (× or ÷) (or Negative, Positive)	× or ÷	Negative

Note: To eliminate (clean up) a double sign, imagine an implied 1 with the sign that stands alone. Using rules 4–6, multiply to combine the two signs. In the first example, clean up the double sign by multiplying the ⁻4 by ⁻1. The result is ⁺4 (Rule 5). In the second example, the implied 1 is positive, so multiply ⁻5 by ⁺1. The result is ⁻5 (Rule 6).

Examples:
⁻3 − (⁻4)   7 + (⁻5)
⁻3 − 1(⁻4)   7 + 1(⁻5)
⁻3 + 4   7 − 5

Example:

$3[2 + 4(3)^2]$ Solve the exponent first.

$3[2 + 4(9)]$ Multiply inside the brackets.

$3[2 + 36]$ Add inside the brackets.

$3[38] = 114$ Multiply.

Algebra Ready — Lesson 9

5 Guided Learning

Definition or Problem	Notes and Explanations
	Use the order of operations and the six signed number rules to solve.
$16 - 18 \div 3(13 - 16)^2$	Work inside the parentheses first.
$16 - 18 \div 3(^-3)^2$	Solve the exponent.
$16 - 18 \div 3(9)$	Divide and multiply left to right.
$16 - 6(9)$	
$16 - 54$	Subtract.
$^-38$	
$3 + 4(5 - 9)$	Work inside the parentheses first.
$3 + 4(^-4)$	Multiply.
$3 + (^-16)$	Eliminate the double sign using the implied 1 and Rule 6.
$3 + 1(^-16)$	
$3 - 16$	Subtract.
$^-13$	
	Follow the order of operations and the six signed number rules to solve.
$^-3 - (^-7) + 4 - 5 - (^-2)$	To clean up the double signs, you must multiply.
$^-3 + 7 + 4 - 5 + 2$	Now add and subtract from left to right.
$4 + 4 - 5 + 2$	
$8 - 5 + 2$	
$3 + 2$	
5	

Guided Class Practice

Let's try these examples.

1. 13 − 22   **-9**

2. -42 + 23(-5)   **-157**

3. -17 + 45   **28**

4. -4 − (-12) − (-3)   **11**

5. (-6)(15) + 13   **-77**

6. 4(5 + 12)   **68**

7. $(2 + 2)^2(5 + 4)$ **144**

8. $18 − (3 \times 3)^2 + 78$ **15**

Algebra Ready — Lesson 9

Name _____

Date _____

Period _____

7 Independent Practice

Simplify the following problems. Show all work and circle your answers.

1. $(^-6) - (^-4)$ **-2**

2. $9 - 4 - (^-10)$ **15**

3. $(^-4)(^-5) - (6)(^-3)$ **38**

4. $^-3 - (^-5) - (^-6) - (^-7)$ **15**

5. $^-3 - (^-3) + 4 - 5 - (^-6)$ **5**

6. $5 - (^-3 - 4) - 9 - (^-6)$ **9**

7. $^-3 - (^-2) - 1 - 3 - (^-5)$ **0**

8. $(^-6 - 3 + 4) - (^-8 - 4 + 2)$ **5**

9. $(^-4 - 6) - (5 - 9)$ **-6**   10. $^-3 - 3(5 - 10)$ **12**

11. $^-4(2)^4$ **-64**   12. $(^-3)(4) + (^-6)(^-2)$ **0**

13. $(^-2)(^-4)(^-5) + (6)(3)(2)(^-1)$ **-76**   14. $^-44 \div 11$ **-4**

15. $12 - 4(^-5 + 3)^4 - 8$ **-60**   16. $^-12 - (^-5) + 6$ **-1**

17. $^-11 - (^-7) + 6$ **2**   18. $^-8(^-6) - 4(^-6)$ **72**

BONUS

$^-16 - 4(7) - (^-12) + 5(^-3)$ **-47**

Algebra Ready — Lesson 10

1. Word Problem Warm-Up

Cindy was given the following challenge by her brother Brian. Brian said, "I'm thinking of two consecutive numbers whose product is 42 and whose sum is 13. What are the two numbers, Cindy?" What answer should Cindy give? __6 & 7__

Show work here.

Problem-solving plan…	
Read	
Key words	
Analyze	
Add	
Subtract	
Multiply	
Divide	
Compare/Check	

2. LESSON 10: OBJECTIVES

1. Review the order of operations using several types of grouping symbols.
2. Practice the six signed number rules.

3. Vocabulary

1. ***consecutive numbers*** — Consecutive numbers are numbers that follow each other in a sequential pattern.
2. **sum** — The sum is the answer to an addition problem.
3. **product** — The product is the answer to a multiplication problem.
4. **difference** — The difference is the answer to a subtraction problem.
5. **quotient** — The quotient is the answer to a division problem.

④ Lesson Presentation

1. Review the order of operations using several types of grouping symbols.
2. Practice the six signed number rules.

Order of Operations

1. Work inside the parentheses or other grouping symbols. When grouping symbols are nested { [( )] }, work from the inside to the outside.
2. Solve the exponent(s).
3. Multiply or divide from left to right.
4. Finally, add or subtract from left to right.

Please **E**xcuse **M**y **D**ear **A**unt **S**ally

Six Signed Number Rules

Complete the following signed number rules by filling in the blanks.

Combination of Numbers	Operation	Answer Looks Like
1. Positive, Positive	+	*Positive*
2. Negative, Negative	+	*Negative*
3. Positive, Negative (or Negative, Positive)	−	*The sign of the "larger" number*
4. Positive, Positive (× or ÷)	× or ÷	*Positive*
5. Negative, Negative (× or ÷)	× or ÷	*Positive*
6. Positive, Negative (× or ÷) (or Negative, Positive)	× or ÷	*Negative*

Example:

$$-8(-3) \div 3 - 21 \div 7 \div (-3) + 6$$
$$24 \div 3 - 21 \div 7 \div (-3) + 6$$
$$8 - 21 \div 7 \div (-3) + 6$$
$$8 - 3 \div (-3) + 6$$

Multiply and divide from left to right.

$$8 + 1 + 6$$

Now add from left to right.

$$9 + 6 = 15$$

Algebra Ready — Lesson 10

5 Guided Learning

Definition or Problem	Notes and Explanations
	Note: Students may need to be reminded that a · means multiply.
2[3(4 · 5 + 6) + 1]	*When you have nested grouping symbols,*
2[3(20 + 6) + 1]	*work from the inside out.*
2[3(26) + 1]	*Now, work within the brackets.*
2[78 + 1]	
2[79]	*Like parentheses, brackets mean multiply.*
158	
(3 + 6 − 5) + 3(6 + 1)	*Work inside the parentheses first.*
4 + 3(7)	*Multiply.*
4 + 21	*Add.*
25	
	Follow the order of operations.
4 ÷ 2 × 1 + 4 ÷ 2	*Divide first.*
2 × 1 + 4 ÷ 2	*Multiply.*
2 + 4 ÷ 2	*Divide.*
2 + 2	*Add.*
4	

Definition or Problem	Notes and Explanations

6 Guided Class Practice

Let's try these examples.

1. $-3[2 - 2(-2 - 2) - 2]$ **-24**

2. $4[5 + 3(2 - 4)^2]$ **68**

3. $-12 \div 6 \cdot 3 - (4 - 2)$ **-8**

4. $(5 - 2)^3 + (4 + 2 + 1)^2$ **76**

Algebra Ready — Lesson 10

Name _____
Date _____
Period _____

7 Independent Practice

Simplify the following problems. Show all work and circle your answers.

1. $-2[-1 - 2(-2 - 2)^2 - 3]$ **72**

2. $-8(-3) \div 6 - 28 \div 7 \div 2 + 6$ **8**

3. $-8 \div 2 \times 5 - (6)(-3)$ **-2**

4. $-2(3 - 6)^2 - 5(-1 - 2)^3$ **117**

5. $-3 \cdot 6 - (-4 - 7)$ **-7**

6. $(-9)^2 - (-8)^2 + (-7)^2 + 8^2$ **130**

7. $-1(-1) \div 1 + (-1) - (-1)(-1)$ **-1**

8. $(-6 - 3 + 4)^2 - (8 - 4 + 2)^2$ **-11**

9. $(^-4 - 6) - (5 - 9)$ **$^-6$**

10. $^-3 \cdot 5 - 10 \cdot 5$ **$^-65$**

11. $\frac{8}{7} - \frac{6}{7}$ **$\frac{2}{7}$**

12. $^-3 \times 4 + (^-6)(^-2) - (^-2)(^-3)$ **$^-6$**

13. $(^-2)(^-3)(^-4)(^-1)(^-2)(^-5)$ **240**

14. $^-64 \div (^-4)$ **16**

15. $^-2(^-8 + 5)^3 - 8 - (^-3)^2$ **37**

16. $14 \div 2 - 3[^-2 - 3(3 - 8)^2 - 4]$ **250**

17. $12(^-3) \div (^-9) + 48 \div (^-6) \div (^-4)$ **6**

18. $^-8(^-6) - 4 \cdot 6$ **24**

BONUS

$25 + [5 - (^-5)]^2 + 4(2 - 5 + 15) - 75$ **98**

Algebra Ready

Try This 8

Words Into Symbols

Write numeric or algebraic expressions for the following.

Note: For simplicity's sake, always use x or n for an unknown number.

1. the product of 2 and 4

   $2 \times 4;\ 2 \cdot 4;\ (2)(4)$

2. the difference between 7 and 3

   $7 - 3$

3. 6 divided by 2

   $6 \div 2;\ 2\overline{)6};\ \frac{6}{2}$

4. the sum of 7 and 2

   $7 + 2$

5. 9 subtracted from 12

   $12 - 9$

6. 4 times 3

   4×3

7. 6 doubled

   6×2

8. 3 to the second power

   3^2

9. the product of a number and 4

   $4n;\ 4 \cdot n$

10. the sum of a number squared and the same number cubed

    $x^2 + x^3$

Write numeric or algebraic expressions for the following.

11. the sum of 3 and 4 over the sum of 6 and 8

$$\frac{3+4}{6+8}$$

12. the quotient of a number and 7

$n \div 7;\ 7\overline{)n};\ \frac{n}{7}$

13. the product of x and y squared

xy^2 or $x \cdot y^2$

14. the sum of a and b over the difference of x and y

$$\frac{a+b}{x-y}$$

15. the product of a squared, b squared, and c squared

$a^2 b^2 c^2$

16. the square of the sum of 5x and 2

$(5x + 2)^2$

17. 30 decreased by x tripled

$30 - 3x$

18. x increased by y

$x + y$

19. the sum of a and 3 decreased by the sum of b and 7

$(a + 3) - (b + 7)$

20. 12 added to negative 3

$^-3 + 12$

21. the product of negative 7 and y divided by 2

$\frac{^-7y}{2}$ or $^-7y \div 2$

22. the sum of x and 3 equals 7

$x + 3 = 7$

Algebra Ready — Lesson 11

1. Word Problem Warm-Up

Last winter, the low temperature one day was ⁻5° F. By noon the temperature had risen by 22° F. What was the temperature at noon?

17° F

The following day, the temperature reached 5° F before dropping 25° F to hit the day's low temperature. What was the low temperature?

⁻20° F

Show work here.

Problem-solving plan...
- Read
- Key words
- Analyze
- Add
- Subtract
- Multiply
- Divide
- Compare/Check

2. LESSON 11: OBJECTIVES

1. Review the order of operations and the signed number rules.
2. Practice addition, subtraction, multiplication, division, and exponents.

3. Vocabulary

1. *numerator* — The numerator is the number above the line in a fraction.
2. *denominator* — The denominator is the number below the line in a fraction.
3. **equation** — An equation is a mathematical sentence that has two equivalent expressions separated by an equal sign.

 **Lesson Presentation**

1. Review the order of operations and the signed number rules.
2. Practice addition, subtraction, multiplication, division, and exponents.

Order of Operations

1. Work inside the grouping symbols. When grouping symbols are nested { [( )] }, work from the inside to the outside. A fraction bar also acts as a grouping symbol.
2. Solve the exponent(s).
3. Multiply or divide from left to right.
4. Finally, add or subtract from left to right.

Six Signed Number Rules

Combination of Numbers	Operation	Answer Looks Like
1. Positive, Positive	+	Positive
2. Negative, Negative	+	Negative
3. Positive, Negative (or Negative, Positive)	−	The sign of the "larger" number
4. Positive, Positive (× or ÷)	× or ÷	Positive
5. Negative, Negative (× or ÷)	× or ÷	Positive
6. Positive, Negative (× or ÷) (or Negative, Positive)	× or ÷	Negative

Note: To clean up a double sign, multiply by the implied 1 and follow rules 4–6 to combine the signs.

Example:

$2^2 + [(6 + 3 \cdot 2) + (^-2)^3]$ Follow the order of operations. Work inside the sets of parentheses.

$2^2 + [(6 + 6) + (^-2)^3]$

$2^2 + [12 + (^-2)^3]$

$2^2 + [12 + (^-8)]$ Combine numbers inside the brackets. Remember the implied 1 to clean up the double sign.

$2^2 + [12 - 8]$ Work inside the brackets.

$2^2 + [4]$ Solve the exponent.

$4 + 4 = 8$ Add.

Algebra Ready — Lesson 11

5. Guided Learning

Definition or Problem	Notes and Explanations
	Integer addition and subtraction may need to be reviewed.
$6 + 5 = 11$	Rule 1
$-3 - 4 = -7$	Rule 2
$3 - 4 = -1$	Rule 3
$-11 + 9 = -2$	Rule 3
	Integer multiplication and division may need to be reviewed.
$3 \cdot 3 = 9$	Rule 4
$(-4)(-2) = 8$	Rule 5
$-2(6) = -12$	Rule 6
$(5)(-3) = -15$	Rule 6
	Note: A fraction bar means to divide and is used as a grouping symbol in order of operations problems.
$\dfrac{-6 + 2(9)}{2 + 8 - 14}$	Simplify the numerator by following the order of operations. Multiply, then add.
$\dfrac{-6 + 18}{2 + 8 - 14}$	
$\dfrac{12}{2 + 8 - 14}$	Simplify the denominator by following the order of operations. Add, then subtract.
$\dfrac{12}{10 - 14}$	
$\dfrac{12}{-4}$	Divide.
-3	

Definition or Problem	Notes and Explanations

6 Guided Class Practice

Let's try these examples.

1. $(-3)^3 + (2)^5$ **5**

2. $32 \div 2 - [^-3(2^2 + 4)]$ **40**

3. $\dfrac{63 - 18}{^-3 + 1 + 7}$ **9**

4. $16 - [^-3(2^2 + 4 \cdot 3)]$ **64**

Algebra Ready

Lesson 11

Name _____

Date _____

Period _____

7 Independent Practice

Simplify the following problems. Show all work and circle your answers.

1. $^-3 - 6 - 7 - 8$ **-24**

2. $^-9 - 7 - 4 + 3 + 8$ **-9**

3. $(^-11)(^-8)$ **88**

4. $\frac{^-144}{12} + \frac{^-36}{^-9}$ **-8**

5. $^-4 - 3 - 4(5 - 10)$ **13**

6. $^-4(^-5 + 3) - 5(^-3 - 6)$ **53**

7. $14 \div 2 - 2(^-5 - 2)^2$ **-91**

8. $(^-5)^3 - (^-3)^2$ **-134**

9. $2^4 - (^-2)^3 + (^-3)^3$ **-3**

10. $\dfrac{^-8 + 15 + 7}{^-4 + 5 - 8}$ **-2**

11. $2^3 \cdot 5^2$ **200**

12. $4 - 2(^-5)^2 - (^-4)^2$ **-62**

13. $^-3[4 - 7(^-3 - 5)] - (^-5)^2$ **-205**

14. $(^-1)^5 + (^-1)^3 - (^-1)^4 - 1$ **-4**

15. $\dfrac{^-3 - (2^4 - 11)^2}{4^2 - (3)^2}$ **-4**

16. $^-3\{3 - 2[10 - 3(^-2)^2]^2\}^2$ **-75**

17. $(^-2)^6 - (^-3)^4 - (^-6)^3$ **199**

18. $^-1[^-2(^-3)^3]^2$ **-2,916**

Algebra Ready

Lesson 11

Name _____

Date _____

Period _____

7. More Independent Practice

Simplify the following problems. Show all work and circle your answers.

1. $8 - 13 - 3(^-8 + 6)$ **1**

2. $7(^-6 + 1) - 6(^-4 - 4)$ **13**

3. $^-4 - 3(^-9 + 5)^3$ **188**

4. $(^-3)^3 - (8)^2$ **-91**

5. $4^2 - (^-3)^2 + (^-4)^3$ **-57**

6. $\dfrac{^-3 - 4 - 5}{^-2 - 2 - 2}$ **2**

7. $(^-4)^3 \cdot 3^3$ **-1,728**

8. $^-12 \div 2 - 3(^-4)^2 - (^-3)^3$ **-27**

9. $-5[1 - 4(3 - 9)] - (-4)^3$ **-61**

10. $\dfrac{5 + 4[3^2 - (3 - 5)]}{-4 - 5 - 7 + 8 + 1}$ **-7**

11. $(-1)^7 - (-1)^6 - (-1)^3 - (-1)^4$ **-2**

12. $[(3 \cdot 3 + 1)(2 + 3)^2]^2$ **62,500**

13. $5^3 - (-4)^3 + (-3)^5$ **-54**

14. $-5[-1(-2)^4]^2$ **-1,280**

15. $-3\{-3 - 3[-3 - 3(2)] - 3\}$ **-63**

16. $2 - 7[12 - 3(-2)^2]^6$ **2**

Algebra Ready — Lesson 12

1. Word Problem Warm-Up

Jared borrowed $3 from his mom and $9 from his dad. He has paid his mom $1 and his dad $4. How much money does he still owe his parents? ___$7___

Cindy, Jared's sister, borrowed $2 from her mom, $3 from her dad, and $5 from Jared. She has paid Jared back $2. How much money does she still owe her family? ___$8___

Show work here.

Problem-solving plan...
Read	
Key words	
Analyze	
Add	
Subtract	
Multiply	
Divide	
Compare/Check	

2. LESSON 12: OBJECTIVE

Solve addition and subtraction equations with variables.

3. Vocabulary

1. **variable** — A variable is a symbol used to take the place of an unknown number.
2. **inverse operations** — Inverse operations are opposite mathematical operations that undo each other.
3. **substitute** (verb) — To substitute means to replace a variable or a symbol with a known numerical value.
4. **algebra** — Algebra is a part of mathematics that deals with variables, symbols, and numbers.
5. **isolate** — To isolate is to get something by itself.

4. Lesson Presentation

> Solve addition and subtraction equations with variables.

Addition and Subtraction Equations With Variables

Algebra explores number relationships using letter symbols (variables) to represent unknown numbers. Solving an algebraic equation means finding the value of the variable.

In an algebraic addition or subtraction equation, a number is added to or subtracted from a variable on one side of the equation. To solve addition or subtraction equations, use the following steps.

1. Identify the variable. The goal is to isolate the variable on one side.
2. To isolate the variable, do the inverse operation to eliminate the number being added to or subtracted from the variable. The inverse operation of addition is subtraction. The inverse operation of subtraction is addition.
3. Do the exact same thing to the other side of the equation to keep the sides equal.

Example 1:

$x + 14 = 12$ Identify the variable. Here, it is the x.

$\textcircled{x} + 14 = 12$ Do the inverse operation of addition. Subtract 14 from both sides of the equation to isolate the variable on one side of the equation.

$$\begin{array}{r} x + \cancel{14} = 12 \\ -\cancel{14} \; -14 \\ \hline x = {}^-2 \end{array}$$

The value of the variable is $^-2$.

Example 2:

$k - 13 = 10$ Identify the variable. Here, it is the k.

$\textcircled{k} - 13 = 10$ Do the inverse operation of subtraction. Add 13 to both sides of the equation to isolate the variable.

$$\begin{array}{r} k - \cancel{13} = 10 \\ +\cancel{13} \; +13 \\ \hline k = 23 \end{array}$$

The value of the variable is 23.

Algebra Ready — Lesson 12

5 Guided Learning

Definition or Problem	Notes and Explanations
	Note: Students may need to review the rules for combining signed numbers.
$x + 10 = {}^-22$	Identify the variable.
$\boxed{x} + 10 = {}^-22$	Do the inverse operation of addition. Subtract 10 from
$x + \cancel{10} = {}^-22$ $\quad\;\; -\cancel{10} \;\; -10$	both sides to isolate the variable.
$x = {}^-32$	
Check	Check by substituting your answer for the variable in the original problem.
$x + 10 = {}^-22$	
${}^-32 + 10 = {}^-22$	
${}^-22 = {}^-22$	
$y - 4 = {}^-3$	Identify the variable.
$\boxed{y} - 4 = {}^-3$	Do the inverse operation of subtraction. Add 4 to both
$y - \cancel{4} = {}^-3$ $\quad\; +\cancel{4} \;\; +4$	sides to isolate the variable.
$y = 1$	
Check	Check by substituting your answer for the variable in the original problem.
$y - 4 = {}^-3$	
$1 - 4 = {}^-3$	
${}^-3 = {}^-3$	

Definition or Problem	Notes and Explanations

6 Guided Class Practice

Let's try these examples.

1. $f + 7 = 13$ $f = 6$

2. $j + 7 = {}^-13$ $j = {}^-20$

3. $t - 24 = 17$ $t = 41$

4. $m - 24 = {}^-17$ $m = 7$

Let's set up an equation for each of the following and solve.

5. A number and 18 is 35.

   $n + 18 = 35$
   $n = 17$

6. A number decreased by 18 is 35.

   $x - 18 = 35$
   $x = 53$

Algebra Ready

Lesson 12

Name _____

Date _____

Period _____

7. Independent Practice

Solve the following. Show all work and circle your answers.

1. $x + 6 = 2$ $x = {-}4$

2. $x + 10 = 18$ $x = 8$

3. $y + 21 = {-}8$ $y = {-}29$

4. $x + 3 = {-}24$ $x = {-}27$

5. $x + 13 = 63$ $x = 50$

6. $x + 29 = {-}44$ $x = {-}73$

7. $y + 6 = {-}9$ $y = {-}15$

8. $x + 15 = {-}11$ $x = {-}26$

9. $w + 18 = {-}15$ $w = {-}33$

10. $x + 4 = 3$ $x = {-}1$

11. y − 11 = ⁻20   *y = ⁻9*   12. x − 12 = ⁻5   *x = 7*

13. n − 2 = 12   *n = 14*   14. x − 25 = ⁻10   *x = 15*

15. x − 5 = 23   *x = 28*   16. x − 10 = 27   *x = 37*

17. y − 4 = 26   *y = 30*   18. b − 9 = ⁻14   *b = ⁻5*

19. m − 15 = ⁻72   *m = ⁻57*   20. x − 7 = 20   *x = 27*

21. x − 11 = 36   *x = 47*   22. y − 3 = ⁻24   *y = ⁻21*

23. x − 11 = ⁻40   *x = ⁻29*   24. x − 17 = 13   *x = 30*

Algebra Ready

Try This 8

Mixed Practice

Solve the following. Show all work and circle your answers.

1. $y + 21 = 16$ **y = -5**

2. $x - 13 = -18$ **x = -5**

3. $n + 5 = -90$ **n = -95**

4. $x - 32 = 100$ **x = 132**

5. $x + 27 = -21$ **x = -48**

6. $x - 7 = 30$ **x = 37**

7. $w - 22 = 6$ **w = 28**

8. $x + 7 = -13$ **x = -20**

9. $m + 3 = 4$ **m = 1**

10. $x - 9 = -7$ **x = 2**

11. $x - 8 = 10$ *x = 18*

12. $x - 2 = {}^-7$ *x = ⁻5*

13. $m - 13 = {}^-19$ *m = ⁻6*

14. $x + 22 = 6$ *x = ⁻16*

15. $x + 24 = 14$ *x = ⁻10*

16. $x + 8 = {}^-16$ *x = ⁻24*

17. $w + 6 = 10$ *w = 4*

18. $x - 2 = {}^-12$ *x = ⁻10*

Write an equation for each of the following problems and then solve.

19. A number increased by 9 is ⁻21.

   x + 9 = ⁻21
   x = ⁻30

20. A number and 15 total 45.

   n + 15 = 45
   n = 30

21. The difference between h and 9 is 55.

   h − 9 = 55
   h = 64

22. Six less than a number is 15.

   x − 6 = 15
   x = 21

Algebra Ready　　　　　　　Lesson 13

1. Word Problem Warm-Up

Joe and Sally are trying to compare the amount of candy each has left. They started out with the same amount. Joe has $\frac{2}{5}$ of his candy left, and Sally has $\frac{3}{8}$ of hers left. Use < or > to show who has more candy left.

　　Joe $\frac{16}{40}$ > Sally $\frac{15}{40}$

Show work here.

_____
_____
_____

Problem-solving plan...
Read	
Key words	
Analyze	
Add	
Subtract	
Multiply	
Divide	
Compare/Check	

2. LESSON 13: OBJECTIVES

1. Solve multiplication equations with variables.
2. Review addition and subtraction equations.
3. Convert improper fractions.

3. Vocabulary

1. **least common denominator (LCD)** — The LCD is the smallest multiple that two or more denominators have in common.

2. *improper fraction* — An improper fraction is a fraction in which the numerator is equal to or larger than the denominator.

3. *inequality* — An inequality is a mathematical sentence in which <, >, ≤, or ≥ is used to compare two unequal values.

4. *mixed number* — A mixed number is a value that includes an integer and a fraction.

5. *coefficient* — The coefficient is the number directly in front of a variable. It is the number by which the variable is multiplied.

4. Lesson Presentation

1. Solve multiplication equations with variables.
2. Review addition and subtraction equations.
3. Convert improper fractions.

Multiplication Equations With a Variable

When a variable has a number directly in front of it, that number is the coefficient of the variable. The variable is being multiplied by the coefficient. An equation that involves a coefficient and variable is a multiplication equation. Use the following steps to solve this type of equation.

1. Identify the variable and the coefficient. The goal is to isolate the variable on one side of the equation.
2. Do the inverse operation of multiplication to eliminate the coefficient. That is, divide by the coefficient.
3. Do the same thing to the other side of the equation to keep the sides equal.

Example:

$-8y = 40$	Identify the variable and the coefficient. Here, y is the variable and -8 is the coefficient.
$-8\textcircled{y} = 40$	Divide both sides of the equation by the coefficient of y, -8. Use the fraction bar to show division.
$\dfrac{-8y}{-8} = \dfrac{40}{-8}$	The coefficient cancels out, isolating the variable.
$y = -5$	The value of the variable is -5.

5. Guided Learning

Definition or Problem	Notes and Explanations
$3y = -27$	Identify the variable and the coefficient.
$3\textcircled{y} = -27$	Undo the multiplication by dividing both sides by the coefficient of y, 3.
$\dfrac{3y}{3} = \dfrac{-27}{3}$	
$y = -9$	Note: The fraction bar is used to show division.
Check	Check by substituting your answer for the variable in the original problem.
$3y = -27$	
$3(-9) = -27$	
$-27 = -27$	

Algebra Ready — Lesson 13

Definition or Problem	Notes and Explanations
$-4x = 33$	Identify the variable. The coefficient of x is -4.
$-4\widehat{x} = 33$	Divide both sides by the coefficient of x.
$\dfrac{-4x}{-4} = \dfrac{33}{-4}$	
$x = \dfrac{33}{-4}$	This answer is an improper fraction.
Convert $\dfrac{33}{-4}$ to a mixed number.	Convert the improper fraction to a mixed number by dividing the numerator by the denominator. The remainder becomes the numerator of the fraction in the mixed number.
$-4 \overline{)33} \quad \dfrac{-8\ R1}{} = -8\dfrac{1}{4}$ -32 1	
$x = -8\dfrac{1}{4}$	Note: Students may need to review converting improper fractions to mixed numbers.
Check	Check by substituting your answer for the variable in the original problem. Use the improper fraction to do this.
$-4x = 33$	
$-4\left(\dfrac{33}{-4}\right) = 33$	Note: The $\cdot$ may also be used to show multiplication $\left(-4 \cdot \dfrac{33}{-4}\right)$.
$\dfrac{-4}{1}\left(\dfrac{33}{-4}\right) = 33$	
$\dfrac{33}{1} = 33$	
$33 = 33$	

Definition or Problem	Notes and Explanations

 Guided Class Practice

Let's try these examples.

1. $-3x = 27$ *x = -9*

2. $5k = -55$ *k = -11*

3. $-2m = -48$ *m = 24*

4. $4n = 21$ $n = \frac{21}{4}$ or $5\frac{1}{4}$

Set up an equation for each of the following and solve.

5. The product of a number and -7 is -28.

   $-7x = -28$

   $x = 4$

6. Twelve times a number is 30.

   $12x = 30$

   $x = \frac{30}{12}$ or $2\frac{1}{2}$

Algebra Ready

Lesson 13

Name _____
Date _____
Period _____

7 Independent Practice

Solve each of the following. Show all work and circle your answers.

1. $^-6m = 24$ *m = -4*

2. $x + 4 = 2$ *x = -2*

3. $8x = 16$ *x = 2*

4. $^-9y = 90$ *y = -10*

5. $^-13x = 39$ *x = -3*

6. $x + 3 = 11$ *x = 8*

7. $y - 2 = 6$ *y = 8*

8. $^-5y = 20$ *y = -4*

9. $y + 5 = 17$ *y = 12*

10. $10x = 30$ *x = 3*

11. y − 13 = ⁻7   *y = 6*

12. 7x = 70   *x = 10*

13. ⁻5x = ⁻60   *x = 12*

14. n − 17 = ⁻10   *n = 7*

15. x + 4 = ⁻11   *x = ⁻15*

16. 11x = 22   *x = 2*

17. 8y = ⁻24   *y = ⁻3*

18. x − 2 = 2   *x = 4*

19. ⁻4x = ⁻32   *x = 8*

20. x + 15 = ⁻10   *x = ⁻25*

21. x − 11 = ⁻3   *x = 8*

22. ⁻2x = 20   *x = ⁻10*

23. 6m = 17   *m = $\frac{17}{6}$ or $2\frac{5}{6}$*

24. ⁻5x = 15   *x = ⁻3*

Algebra Ready — Lesson 13

Name _____
Date _____
Period _____

7. More Independent Practice

Solve each of the following. Show all work and circle your answers.

1. $^-10n = 3$ $n = -\frac{3}{10}$

2. $6m = ^-48$ $m = ^-8$

3. $^-12x = ^-12$ $x = 1$

4. $7x = ^-14$ $x = ^-2$

5. $x - 17 = ^-37$ $x = ^-20$

6. $5x = 75$ $x = 15$

7. $y - 9 = ^-13$ $y = ^-4$

8. $^-4x = 39$ $x = \frac{39}{-4}$ or $^-9\frac{3}{4}$

9. $^-9w = 9$ $w = ^-1$

10. $8x = ^-4$ $x = -\frac{1}{2}$

11. $^-7x = 700$ $x = -100$

12. $w + 2 = {}^-3$ $w = -5$

13. $^-6x = {}^-13$ $x = \frac{13}{6}$ or $2\frac{1}{6}$

14. $3w = 21$ $w = 7$

15. $p - 26 = {}^-38$ $p = -12$

16. $^-7x = {}^-2$ $x = \frac{2}{7}$

17. $x + 13 = 13$ $x = 0$

18. $15w = 3$ $w = \frac{1}{5}$

Write an equation for each of the following problems and then solve.

19. A number and 35 total $^-32$.

$x + 35 = -32$
$x = -67$

20. The difference between w and 19 is $^-5$.

$w - 19 = -5$
$w = 14$

21. Fifteen times a number is 90.

$15x = 90$
$x = 6$

22. Fourteen less than a number is $^-2$.

$n - 14 = -2$
$n = 12$

Algebra Ready — Lesson 14

Word Problem Warm-Up

Jose, Candice, and Eddie want to compare their test scores on a recent math test. Jose got $\frac{9}{10}$ of the answers correct, Candice got $\frac{4}{5}$ of the answers right, and Eddie, who was sick and missed the week before the test, answered only $\frac{2}{6}$ of the questions correctly. Order the students' scores from least to greatest. Make sure to prove your answer by showing the necessary work.

Eddie $\frac{2}{6}$ Candice $\frac{4}{5}$ Jose $\frac{9}{10}$

Show work here.

Problem-solving plan...
- Read
- Key words
- Analyze
- Add
- Subtract
- Multiply
- Divide
- Compare/Check

LESSON 14: OBJECTIVES

1. Solve two-step equations.
2. Reduce fractions.
3. Practice addition, subtraction, multiplication, and division with integers.

Vocabulary

1. *triple* — To triple a number means to multiply it by 3.
2. *factor* (noun) — A factor is a number that divides evenly into another number.
3. *greatest common factor* (GCF) — The GCF is the largest factor common to two or more numbers.
4. *unlike terms* — Unlike terms are terms that do not have identical variables.

4. Lesson Presentation

1. Solve two-step equations.
2. Reduce fractions.
3. Practice addition, subtraction, multiplication, and division with integers.

Two-Step Equations

A two-step equation is a combination of an addition or subtraction equation and a multiplication equation. To solve a two-step equation, follow these steps.

1. Identify the variable. The goal is to isolate the variable on one side of the equation.
2. To isolate the variable, do the inverse operation to undo the addition or subtraction.
3. Do the exact same thing to the other side of the equation to keep the sides equal.
4. To solve the remaining multiplication equation, divide both sides by the coefficient of the variable.

Example:

$2x - 5 = {}^-9$	Identify the variable.
$2\underline{x} - 5 = {}^-9$	Do the inverse operation to undo the subtraction. Add 5 to both sides to keep the sides of the equation equal.
$\begin{aligned}2x - 5 &= {}^-9 \\ +5 &+ 5\end{aligned}$	
$2x - 0 = {}^-4$	Divide both sides by 2, the coefficient of x. Remember, the fraction bar shows division.
$\dfrac{2x}{2} = \dfrac{{}^-4}{2}$	
$x = {}^-2$	The value of x is ${}^-2$.

5. Guided Learning

Definition or Problem	Notes and Explanations
${}^-3x + 4 = 16$	*Identify the variable.*
${}^-3\underline{x} + 4 = 16$	*First, undo the addition by subtracting 4 from both sides.*
$\begin{aligned}{}^-3x + 4 &= 16 \\ -4 & -4\end{aligned}$	
${}^-3x = 12$	*Now undo the multiplication by dividing both sides by the coefficient of x, ${}^-3$.*
$\dfrac{{}^-3x}{{}^-3} = \dfrac{12}{{}^-3}$	
$x = {}^-4$	

Algebra Ready — Lesson 14

Definition or Problem	Notes and Explanations
Check	*Check by substituting your answer for the variable in*
$-3x + 4 = 16$	*the original problem.*
$-3(-4) + 4 = 16$	
$12 + 4 = 16$	
$16 = 16$	
$6x - 3 = -24$	*Identify the variable.*
$6\boxed{x} - 3 = -24$	*Undo the subtraction with addition.*
$6x - 3 = -24$	*Undo the multiplication by dividing by the coefficient of x.*
$\quad + 3 \;\; + 3$	*Convert the improper fraction. Reduce fraction by GCF.*
$6x = -21$	
$\dfrac{6x}{6} = \dfrac{-21}{6}$	
$x = -\dfrac{21}{6} = -3\dfrac{3}{6}$ or $-3\dfrac{1}{2}$	
Check	*Check by substituting your answer for the variable in the*
$6x - 3 = -24$	*original problem. Use the improper fraction.*
$6\left(-\dfrac{21}{6}\right) - 3 = -24$	
$-21 - 3 = -24$	
$-24 = -24$	
Review: Finding GCFs	*Note: Students may need review reducing fractions and finding GCFs.*
List the factors for 4 and 24.	*Example: Reduce $\dfrac{4}{24}$.*
4 1 2 2 ④	
24 1 2 3 ④ 6 8 12 24	
4 is the GCF.	

Definition or Problem	Notes and Explanations
Review: Finding GCFs	The GCF can also be found with factor trees.
Factor trees for $\frac{4}{24}$.	The greatest common factor is the product of the prime factors that are common to both numbers. Branch out your factor tree until all prime factors are found. Match up the prime factors common to both numbers. The factors that are common are 2 and 2.
$2 \cdot 2 = 4$	Multiply these common factors. 4 is the GCF.
$\frac{4 \div 4}{24 \div 4} = \frac{1}{6}$	Divide the numerator and denominator by the GCF to reduce the fraction.

⑥ Guided Class Practice

Let's try these examples.

1. $-3x + 6 = 27$ $x = -7$

2. $5k + 20 = -55$ $k = -15$

3. $-2m - 60 = -48$ $m = -6$

4. $4n - 14 = 18$ $n = 8$

Algebra Ready

Lesson 14

Name _____

Date _____

Period _____

7. Independent Practice

Solve each of the following. Show all work and circle your answers.

1. $x + 6 = 11$ **x = 5**

2. $x - 8 = 15$ **x = 23**

3. $n + 4 = {}^-5$ **n = -9**

4. $w - 8 = {}^-6$ **w = 2**

5. $x + 11 = 7$ **x = -4**

6. $m - 16 = {}^-5$ **m = 11**

7. $2y = 8$ **y = 4**

8. $3w = 15$ **w = 5**

9. ${}^-5x = {}^-30$ **x = 6**

10. ${}^-11n = 33$ **n = -3**

11. $^-5y = 11$ $y = -\frac{11}{5} \text{ or } -2\frac{1}{5}$ 12. $6x = {}^-42$ $x = {}^-7$

13. $3m + 4 = 28$ $m = 8$ 14. $2w + 16 = 30$ $w = 7$

15. $5y - 15 = 10$ $y = 5$ 16. $5w + 11 = 31$ $w = 4$

17. $7x + 3 = 32$ $x = \frac{29}{7} \text{ or } 4\frac{1}{7}$ 18. $24y - 9 = 15$ $y = 1$

19. $7w + 10 = {}^-4$ $w = {}^-2$ 20. $5m + 5 = 45$ $m = 8$

Set up an equation for the following problems and solve.

21. A number tripled and 6 is $^-24$.
    $3x + 6 = {}^-24$
    $x = {}^-10$

22. Three less than a number doubled is 15.
    $2x - 3 = 15$
    $x = 9$

Algebra Ready
Lesson 14

Name _____

Date _____

Period _____

7 More Independent Practice

Solve each of the following. Show all work and circle your answers.

1. $^-3x - 11 = 31$ *x = -14*

2. $^-43n + 43 = 43$ *n = 0*

3. $^-4m + 4 = ^-36$ *m = 10*

4. $^-7x + 5 = ^-16$ *x = 3*

5. $^-4w + 7 = 39$ *w = -8*

6. $9x + 4 = ^-32$ *x = -4*

7. $5x - 8 = ^-13$ *x = -1*

8. $^-3w + 4 = ^-22$ *w = $\frac{26}{3}$ or $8\frac{2}{3}$*

9. $4x + 6 = ^-14$ *x = -5*

10. $7y - 11 = ^-35$ *y = $-\frac{24}{7}$ or $-3\frac{3}{7}$*

Write an equation for each of the following problems and then solve.

11. A number increased by 100 is 100.

$$x + 100 = 100$$
$$x = 0$$

12. A number and 16 total ⁻100.

$$n + 16 = {^-}100$$
$$n = {^-}116$$

13. The product of ⁻2 and x is ⁻100.

$$^-2x = {^-}100$$
$$x = 50$$

14. The difference between w and 19 is ⁻100.

$$w - 19 = {^-}100$$
$$w = {^-}81$$

BONUS

15. Seven more than 5 times a number is 57.

$$5x + 7 = 57$$
$$x = 10$$

16. Four less than a number doubled is ⁻12.

$$2x - 4 = {^-}12$$
$$x = {^-}4$$

17. Write a Word Problem Warm-Up that involves decimals.

Answers will vary.

18. Solve #17.

Answers will vary.

Algebra Ready — Lesson 15

1. Word Problem Warm-Up

Bob, Jane, and Hank all invest in the stock market. The price of Bob's stock has risen by $150.00, Jane's stock has risen by $286.00, and Hank's stock has gone up by $135.35. What is the average amount their stocks have increased in value? __$190.45__

Show work here.

Problem-solving plan...
Read	
Key words	
Analyze	
Add	
Subtract	
Multiply	
Divide	
Compare/Check	

2. LESSON 15: OBJECTIVES

1. Review two-step equations.
2. Practice addition, subtraction, multiplication, and division with integers and fractions.

3. Vocabulary

1. **double** — To double a number means to multiply it by 2.
2. **thousands place** — The thousands place is fourth to the left of a decimal point or the fourth from the last digit of a whole number (8,297.64).
3. **average** — An average is the sum of the values of a set of items divided by the number of items added.
4. **mean** — The mean is another word (synonym) for average.

4. Lesson Presentation

1. Review two-step equations.
2. Practice addition, subtraction, multiplication, and division with integers and fractions.

Two-Step Equations

1. Identify the variable. The goal is to isolate the variable.
2. Undo the addition or subtraction by performing the inverse operation.
3. Solve the remaining multiplication equation.

Example:

$^-4x + 7 = ^-9$	Identify the variable.
$^-4\boxed{x} + 7 = ^-9$	Undo the addition by subtracting 7 from both sides.
$^-4x + \cancel{7} = ^-9$ $ - \cancel{7}\ -7$	
$^-4x = ^-16$	Undo the multiplication by dividing both sides by the coefficient of x.
$\dfrac{\cancel{^-4}x}{\cancel{^-4}} = \dfrac{^-16}{^-4}$	
$x = 4$	

5. Guided Learning

Definition or Problem	Notes and Explanations
$6x + 9 = ^-3$	Identify the variable.
$6\boxed{x} + 9 = ^-3$	Subtract 9 from both sides.
$6x + \cancel{9} = ^-3$ $ -\cancel{9}\ -9$	
$6x = ^-12$	Divide both sides by 6, the coefficient of x.
$\dfrac{\cancel{6}x}{\cancel{6}} = \dfrac{^-12}{6}$	
$x = ^-2$	

Algebra Ready
Lesson 15

Definition or Problem	Notes and Explanations
Check	**Check by substituting your answer for the variable in the original problem.**
$6x + 9 = -3$	
$6(-2) + 9 = -3$	**Follow the order of operations.**
$-12 + 9 = -3$	
$-3 = -3$	
$7 - 2x = 21$	**Identify the variable.**
$7 - 2\text{ⓧ} = 21$	**Subtract 7 from both sides.**
$\begin{array}{r} 7 - 2x = 21 \\ -7 -7 \end{array}$	
$-2x = 14$	**Divide both sides by -2, the coefficient of x.**
$\dfrac{-2x}{-2} = \dfrac{14}{-2}$	
$x = -7$	
Check	**Check using substitution.**
$7 - 2x = 21$	**Follow the order of operations.**
$7 - 2(-7) = 21$	
$7 + 14 = 21$	
$21 = 21$	

-107-

Definition or Problem	Notes and Explanations

 Guided Class Practice

Let's try these examples.

1. ⁻4x + 3 = 27   *x = ⁻6*

2. 7 = ⁻5 – 6x   *x = ⁻2*

3. ⁻2m – 115 = 115   *m = ⁻115*

4. ⁻6n – 14 = 18   *n = ⁻5 1/3*

Set up an equation for the following problems and solve.

5. The sum of a number doubled and 37 is 61.

    2x + 37 = 61

    x = 12

6. The difference of a number tripled and 4 is ⁻25.

    3x – 4 = ⁻25

    x = ⁻7

Algebra Ready — Lesson 15

Name _____

Date _____

Period _____

⑦ Independent Practice

Solve each of the following. Show all work and circle your answers.

1. $3x + 6 = 12$ **x = 2**

2. $^-4 = 2x - 6$ **x = 1**

3. $^-5x - 8 = 32$ **x = -8**

4. $3x - 5 = 25$ **x = 10**

5. $5x + 6 = 11$ **x = 1**

6. $7y - 8 = 27$ **y = 5**

7. $^-9 = ^-4 - 5x$ **x = 1**

8. $7y + 8 = 13$ **y = $\frac{5}{7}$**

9. $16x - 3 = 5$ **x = $\frac{1}{2}$**

10. $3x + 9 = ^-6$ **x = -5**

11. $9x + 10 = {}^-8$ **x = ⁻2**   **12.** $3n - 15 = {}^-3$ **n = 4**

13. $7x + 6 = {}^-8$ **x = ⁻2**   **14.** $4x - 11 = {}^-5$ **x = 1½**

15. ${}^-3y + 6 = {}^-3$ **y = 3**   **16.** ${}^-2x - 5 = {}^-6$ **x = ½**

17. ${}^-8x + 6 = {}^-4$ **x = 1¼**   **18.** ${}^-16 = 5 + 3w$ **w = ⁻7**

19. ${}^-5w + 9 = {}^-31$ **w = 8**   **20.** ${}^-6x - 3 = 9$ **x = ⁻2**

21. ${}^-4x + 1 = 1$ **x = 0**   **22.** $8 - 5y = 4$ **y = ⅘**

Algebra Ready — Lesson 15

Name _____

Date _____

Period _____

7 More Independent Practice

Solve each of the following. Show all work and circle your answers.

1. $4m + 12 = 24$ m = 3

2. $8 = {}^-8 - 6x$ x = $-2\frac{2}{3}$

3. $^-5x - 11 = {}^-11$ x = 0

4. $6x + 4 = 16$ x = 2

5. $^-8 = 28 + 12m$ m = -3

6. $7k + 15 = {}^-55$ k = -10

7. $3w + 5 = {}^-4$ w = -3

8. $8x - 1 = {}^-49$ x = -6

9. Billy Bob has four times as many marbles as Joey. Sarah then gives Billy Bob 7 more marbles. If Billy Bob now has 43 marbles, how many marbles does Joey have?

   4x + 7 = 43
   x = 9

REVIEW

Place < or > on the lines provided to compare the values of the following pairs of numbers. Remember that < means **less than** and > means **greater than**.

1. −15 __<__ −12

2. 15 __>__ 12

3. −3 __<__ 3

4. −3 __<__ 0

5. −14 __<__ 10

Round **45,382** to the named place value.

6. tens place __45,380__

7. thousands place __45,000__

8. hundreds place __45,400__

9. Find the value of 3^2. __9__

Simplify using the correct order of operations.

10. $2(4 - 9 \div 3)$ __2__

11. $(-21/3 - 100/5) \div (-3)$ __9__

12. $6 \div 2(4 \cdot 3 \div 12)$ __3__

13. $(-1 \cdot -5) + (70 \div 10)/7$ __6__

Algebra Ready — Lesson 16

1. Word Problem Warm-Up

Erin bowled five games. Her scores were 119, 124, 118, 129, and 165. What was Erin's average score? *Erin's average was 131.*

Show work here.

Problem-solving plan...
- Read
- Key words
- Analyze
- Add
- Subtract
- Multiply
- Divide
- Compare/Check

2. LESSON 16: OBJECTIVES

1. Simplify expressions by combining like terms.
2. Add and subtract with integers.

3. Vocabulary

1. **term** — A term is a part of an expression. Terms are separated by addition and subtraction symbols.

2. **coefficient** — A coefficient is the number in front of a variable that tells how many of the variable you have.

3. **like terms** — Like terms are terms that have identical variables.

4. **simplify** — To simplify is to combine like terms and put an answer into its lowest form.

5. **polynomial** — A polynomial is a sum of monomials (for example, $4y + 3$, $6 - 2y + 4z$, $5m^2y + 8my^2$).

6. **binomial** — A binomial is a polynomial with exactly two unlike terms.

Lesson Presentation

1. Simplify expressions by combining like terms.
2. Add and subtract with integers.

Combine Like Terms

Sometimes an expression has several terms. To put this type of expression into simplest form, combine the like terms (terms that have identical variables) following the signed number rules.

1. Look at the variables in the expression. Identify the variable in the first term. Look through the problem for other terms with the identical variable or variable group.
2. Combine the coefficients of these like terms using the signed number rules.
3. Repeat these steps with each new variable term until all like terms are combined.

Example:

$4z + 2x - 6z + 7x$	Look at the variables in this expression. There are z terms and x terms.
$4z + 2x \underline{- 6z} + 7x$	The first term is a z term. Find the other z terms, and combine all z terms.
$^-2z + 2x + 7x$	
$^-2z \underline{+ 2x} \underline{+ 7x}$	Identify and combine the x terms.
$^-2z + 9x$ **or** $9x - 2z$	The original expression has been simplified to a binomial. Usually the terms of an answer are put in alphabetical order by variable; however, the answers are equivalent and therefore both are correct.

Guided Learning

Definition or Problem	Notes and Explanations

Algebra Ready — Lesson 16

Definition or Problem	Notes and Explanations
$4x^2 - 3x^3 - 4x^3 + 2x^2$	
$\underline{4x^2} - 3x^3 - 4x^3 \underline{+ 2x^2}$	Find the like terms of $4x^2$ and combine them.
$6x^2 \underline{- 3x^3} \underline{- 4x^3}$	Find the like terms of $^-3x^3$ and combine them.
$6x^2 - 7x^3$	
$^-7x^3 + 6x^2$	Generally, answers are given with exponents in descending order. In the absence of exponents, alphabetical order determines the order of the terms.
$4ac + 2ab + 7ab + 9ab$	
$\underline{4ac} + 2ab + 7ab + 9ab$	Find the like terms of $4ac$. There are none.
$4ac + \underline{2ab} + \underline{7ab} + \underline{9ab}$	Find the like terms of $2ab$ and combine them.
$4ac + 18ab$ or	
$18ab + 4ac$	Generally, answers are put in alphabetical order; however, both answers are correct.

Definition or Problem	Notes and Explanations

6 Guided Class Practice

Let's try these examples.

1. $-7h - 12y - 10y + 12h$ **5h − 22y**

2. $15a^2 + 12b - 9a^2 - 7a$ **6a² − 7a + 12b**

3. $7y + 12w - 7x + 8w$ **20w − 7x + 7y**

4. $2x - 4x^2 + 8x - x^2 - 4xy$ **-5x² + 10x − 4xy**

Algebra Ready — Lesson 16

Name _____
Date _____
Period _____

Independent Practice

Simplify by combining like terms. Show all work and circle your answers. When you are finished, underline all the answers that are binomials.

1. $5x + 6x + 3x$ **14x**
2. $2m + 5m + 4m$ **11m**

3. $3a + a + 12a$ **16a**
4. $5b + 7b + 4b - 10b$ **6b**

5. $8 + 3s + 8s$ **11s + 8**
6. $4k - 3 + 5k$ **9k - 3**

7. $6x + 5 + x + 2x + 9$ **9x + 14**
8. $5n + 6r + 4n + 11r$ **9n + 17r**

9. $9x - 4y - 6y + 8x - 2y - 3x$ **14x - 12y**
10. $^-8ns + 3ns + 6b + 5b$ **11b - 5ns**

11. $^-3x^2 + x^2 + y^2 + 6y^2$

$\underline{\quad -2x^2 + 7y^2 \quad}$

12. $5a^2 - 3b$

$\underline{\quad 5a^2 - 3b \quad}$

13. $12a - 10b + 3a + 7b$

$\underline{\quad 15a - 3b \quad}$

14. $3 + r^2 + 4r - 2r - r^2$

$\underline{\quad 2r + 3 \quad}$

15. $10 - 3t + 2s + 2$

$\underline{\quad 2s - 3t + 12 \quad}$

16. $2rs - 3 - 5rs + 6rs - s$

$\underline{\quad 3rs - s - 3 \quad}$

17. $9a - 6b - 7c - 4c + 2b$

$\underline{\quad 9a - 4b - 11c \quad}$

18. $5x^3 - 6y + 7x^2 - 9y - 4x^2$

$\underline{\quad 5x^3 + 3x^2 - 15y \quad}$

19. $4am - 7an + 8am + 3an - 4$

$\underline{\quad 12am - 4an - 4 \quad}$

20. $8ab - 5ac + 9ab + 7c$

$\underline{\quad 17ab - 5ac + 7c \quad}$

21. $3ab + 6xy - 5ab + 6y$

$\underline{\quad ^-2ab + 6xy + 6y \quad}$

22. $5c - 4x - 8y - 9x + 3y - 2c$

$\underline{\quad 3c - 13x - 5y \quad}$

Algebra Ready — Lesson 17

1. Word Problem Warm-Up

Matthew earned $5.25 per hour working at Fanny's Flag Shop. One weekend he worked 11 hours. He was hoping to earn enough money that weekend to buy a football that cost $46.00 and a sweatshirt that cost $10.25.

Did he earn enough for both? If he did earn enough, how much money will he have left? Or, if he did not earn enough, how much more will he need? __*Yes, he earned $57.75, so he will have $1.50 left.*__

Show work here.

Problem-solving plan...

Read	
Key words	
Analyze	
Add	
Subtract	
Multiply	
Divide	
Compare/Check	

2. LESSON 17: OBJECTIVES

1. Simplify expressions by combining like terms.
2. Add and subtract with integers.

3. Vocabulary

1. **monomial** — A monomial is an algebraic expression with exactly one term.
2. *trinomial* — A trinomial is a polynomial with exactly three unlike terms.
3. **symbol** — A symbol is a picture or shape that stands for an operation or a constant value.

4. Lesson Presentation

1. Simplify expressions by combining like terms.
2. Add and subtract with integers.

Combine Like Terms

1. Identify the variable in the first term. Look through the problem for other terms with the identical variable or variable group.
2. Combine the coefficients of these like terms.
3. Repeat these steps with each new term until the expression is completely simplified.

Example:

$4y^2 + 2x + 4 + 2 - 6y^2 + 7x$	Look at the variables in this expression. There are y^2 terms and x terms and number terms.
$\underline{4y^2} + 2x + 4 + 2 \underline{- 6y^2} + 7x$	Identify and combine the y^2 terms.
$^-2y^2 + 2x + 4 + 2 + 7x$	
$^-2y^2 \underline{+ 2x} + 4 + 2 \underline{+ 7x}$	Identify and combine the x terms.
$^-2y^2 + 9x + 4 + 2$	Identify and combine the number terms.
$^-2y^2 + 9x + 6$	The original expression has been simplified to a trinomial. Leave the answer in descending order of exponents. Putting the answer in alphabetical order of variables is a secondary consideration when exponents appear in the answer.

5. Guided Learning

Definition or Problem	Notes and Explanations
$^-4x^3 + 2a^2 + 2z^5 + a^2$	
$\underline{^-4x^3} + 2a^2 + 2z^5 + a^2$	Find like terms of $^-4x^3$ and combine them. There are none.
$^-4x^3 + \underline{2a^2} + 2z^5 + \underline{a^2}$	Find like terms of $2a^2$ and combine them.
$^-4x^3 + 3a^2 + \underline{2z^5}$	There are no like terms of $2z^5$.
$2z^5 - 4x^3 + 3a^2$	Rearrange the terms in descending order of exponents.

Algebra Ready — Lesson 17

Definition or Problem	Notes and Explanations
$3x^2y + 2xy^2 - 7x^2y + 10xy^2$	
$\underline{3x^2y} + 2xy^2 \underline{- 7x^2y} + 10xy^2$	$3x^2y$ and $^-7x^2y$ are like terms, so combine them.
$^-4x^2y \underline{+ 2xy^2 + 10xy^2}$	Combine the xy^2 terms.
$^-4x^2y + 12xy^2$	Note: $^-4x^2y$ and $12xy^2$ are not like terms because the variable groups are different. In $^-4x^2y$, the exponent 2 is on the x. In $12xy^2$, the exponent 2 is on the y. Variables of like terms must be identical in every way.
$2ab + 4ac + 2bc + 3ac + 2ab + 4bc$	Note: Notice how similar the variable groups are. Be careful when identifying the like terms.
$\underline{2ab} + 4ac + 2bc + 3ac + \underline{2ab} + 4bc$	Find like terms of $2ab$ and combine them.
$4ab + \underline{4ac} + 2bc + \underline{3ac} + 4bc$	Find like terms of $4ac$ and combine them.
$4ab + 7ac + \underline{2bc} + \underline{4bc}$	Find like terms of $2bc$ and combine them.
$4ab + 7ac + 6bc$	

Definition or Problem	Notes and Explanations

6 Guided Class Practice

Let's try these examples.

1. $-7h^2 - 12y - 10y + 12h^2 + a$

   $5h^2 + a - 22y$

2. $15a^2 + 12b - 9a^2 - 7a + y^3$

   $y^3 + 6a^2 - 7a + 12b$

3. $12y + 12w - 12x - 12w$

   $-12x + 12y$

4. $12x - 4x^2 + 8x^2 - 3x^2 - 4x^2y$

   $-4x^2y + x^2 + 12x$

Algebra Ready — Lesson 17

Name _____
Date _____
Period _____

 Independent Practice

Simplify the following expressions. Show all work and circle your answers. Underline all answers that are trinomials.

1. $4x + 2x + 3x + 5x$

   $14x$

2. $x^2 + y + y^2 + x$

   $x^2 + y^2 + x + y$

3. $x^3 + 2x + 3x^2 - 4x + 5x^3$

   $6x^3 + 3x^2 - 2x$

4. $2a^2 - 4a^2 + 5a^4 - a^2 + a^3 + 6$

   $5a^4 + a^3 - 3a^2 + 6$

5. $3ab + 2b + 6a + 5 + 3b^2 + a$

   $3b^2 + 3ab + 7a + 2b + 5$

6. $7ax + 2x + 6a^2 + 9 - 5x^2 - 9x$

   $6a^2 - 5x^2 + 7ax - 7x + 9$

7. $a^4 + a^6 + a^3 + a^2 + 1 + a^5 + b$

   $a^6 + a^5 + a^4 + a^3 + a^2 + b + 1$

8. $3ab + 6ab^2 + 3a^2b + 3ab^2 + 6ab$

   $3a^2b + 9ab^2 + 9ab$

9. $5b^2 + 6ab + 3a^2 + 4a + 5b + 6$

$3a^2 + 5b^2 + 6ab + 4a + 5b + 6$

10. $8x^3 + 9x - 8x^2 - 7x^2 - 11x - 8x$

$8x^3 - 15x^2 - 10x$

11. $8 + 3a - 6ab + 4a^2 + 2a - 3ab + 9$

$4a^2 + 5a - 9ab + 17$

12. $^-11a^3 + 3a - 4a^3 + a + 6 - 9a^2$

$^-15a^3 - 9a^2 + 4a + 6$

13. $4x^2y - 6xy^2 + 8y - 9x^2y + 5xy^2$

$^-5x^2y - xy^2 + 8y$

14. $5ac - 6ab + 9a - 8ac - 5ab$

$^-11ab - 3ac + 9a$

15. $4x^2 - 5y + 6x^2 - 4y - 3c + 9c$

$10x^2 + 6c - 9y$

16. $2a - c - b - 4c - 3b - 7a + 8c$

$^-5a - 4b + 3c$

17. One problem from this Independent Practice has a seven-term polynomial for its answer. Write that answer.

(7.) $a^6 + a^5 + a^4 + a^3 + a^2 + b + 1$

18. Write your own problem similar to the ones in this assignment. Make sure your problem has at least six terms.

Answers will vary.

19. Simplify the problem you wrote.

Answers will vary.

Algebra Ready — Lesson 18

1. Word Problem Warm-Up

Sandra received a fractional score of $\frac{8}{10}$ on her quiz. What is her score as a percent? __**80%**__

Jim received 95% on his quiz. What is his score as a fraction in the simplest form (reduced)? __$\frac{95}{100}$ **reduced to** $\frac{19}{20}$__

Show work here.

Problem-solving plan...
- Read
- Key words
- Analyze
- Add
- Subtract
- Multiply
- Divide
- Compare/Check

2. LESSON 18: OBJECTIVES

1. Solve two-step equations with variable terms on both sides.
2. Practice the four basic operations with integers and fractions.

3. Vocabulary

1. ***percent*** — Percent is a way to represent part of a whole that has been divided into 100 equal pieces.

2. ***divisible*** — Divisible describes a number that can be divided by another number evenly with no remainder.

3. ***volume*** — Volume is the amount of space inside a three-dimensional object.

4. ***variable term*** — A variable term is a mathematical term with at least one variable.

4 Lesson Presentation

1. Solve two-step equations with variable terms on both sides.
2. Practice the four basic operations with integers and fractions.

Two-Step Equations With Variable Terms on Both Sides

A two-step equation is a combination of an addition or subtraction equation and a multiplication equation. Sometimes a two-step equation has a variable on both sides of the equal sign. In such an equation, the variable terms must be combined on one side of the equation by using the inverse operation. To solve a problem like this, use the following steps.

1. Identify the like terms.
2. Combine the like terms on one side of the equation by using the inverse operation.
3. To solve the remaining multiplication equation, divide both sides by the coefficient of the variable.

Example:

$^{-}5x - 48 = 19x$ There are variable x terms on both sides of the equation. Combine these like terms by adding 5x to both sides.

$\begin{array}{r} ^{-}\cancel{5}x - 48 = 19x \\ + 5\cancel{x} + 5x \\ \hline \end{array}$

$^{-}48 = 24x$ Divide both sides of the equation by 24, the coefficient of x.

Remember: The fraction bar shows division.

$\dfrac{^{-}48}{24} = \dfrac{\cancel{24}x}{\cancel{24}}$

$^{-}2 = x$

$x = ^{-}2$

5 Guided Learning

Definition or Problem	Notes and Explanations

·126·

Algebra Ready   Lesson 18

Definition or Problem	Notes and Explanations
	Note: Make sure students understand that all the variable terms go together on one side.
$2x - 13 = {}^-37x$	**Combine the x terms by subtracting 2x from both sides.**
$\begin{array}{r} 2x - 13 = {}^-37x \\ -2x \qquad\quad -2x \end{array}$	
${}^-13 = {}^-39x$	**Solve the remaining multiplication equation by dividing both sides by the coefficient of x.**
$\dfrac{-13}{-39} = \dfrac{-39x}{-39}$	
$\dfrac{-13}{-39} = x$	**Reduce the fraction. The GCF is $^-13$.**
$\dfrac{-13}{-39} \div \dfrac{-13}{-13} = \dfrac{1}{3}$	
$x = \dfrac{1}{3}$	
$15x = {}^-4 - 9x$	**Combine like terms by adding 9x to both sides.**
$\begin{array}{r} 15x = {}^-4 - 9x \\ +9x \qquad\quad +9x \end{array}$	
$24x = {}^-4$	**Solve by dividing both sides by the coefficient of x.**
$\dfrac{24x}{24} = \dfrac{-4}{24}$	
$x = -\dfrac{4}{24} \div \dfrac{4}{4}$	**Reduce the fraction using the GCF of 4 and 24.**
$x = -\dfrac{1}{6}$	*Note: Remind students that when either the numerator or the denominator of a fraction is negative, the entire fraction is negative.*
	Note: Students may need a review on finding GCFs.

Definition or Problem	Notes and Explanations

6 Guided Class Practice

Let's try these examples.

1. $^-4x + 33 = 7x$ *x = 3*

2. $2k = 7k + 15$ *k = -3*

3. $8x - 18 = 6$ *x = 3*

4. $^-13m = {}^-14 - 6m$ *m = 2*

5. $18x = {}^-100 - 2x$ *x = -5*

6. $4 + 12c = {}^-5$ *c = -$\frac{3}{4}$*

Algebra Ready

Lesson 18

Name _____

Date _____

Period _____

7 Independent Practice

Solve the following equations. Show all work and circle your answers.

1. $-x = 25 - 6x$ **x = 5**

2. $9 - 6x = 3$ **x = 1**

3. $5 - 2m = 13m$ **m = $\frac{1}{3}$**

4. $-3 - 16n = -5$ **n = $\frac{1}{8}$**

5. $-7 + 8x = -39$ **x = -4**

6. $8x + 15 = 3x$ **x = -3**

7. $-2x - 1 = -4x$ **x = $\frac{1}{2}$**

8. $4x + 9 = 13x$ **x = 1**

9. $-3y = 11y + 7$ **y = -$\frac{1}{2}$**

10. $16x = 2x - 42$ **x = -3**

11. $-19x = -7x - 24$ **x = 2**

12. $-3x = -8x + 15$ **x = 3**

13. $16 + 13y = -10$ **y = -2**

14. $-8 - 3w = -14$ **w = 2**

REVIEW

On the lines below, write the steps to follow in the order of operations.

1. *Parentheses and other grouping symbols*
2. *Exponents*
3. *Multiplication or division left to right*
4. *Addition or subtraction left to right*

Simplify the following problems using the order of operations.

1. $-6 - 2(2^4 - 7)^2$ **-168**

2. $8 + 7(2) - (-3)^3$ **49**

Algebra Ready
Lesson 18

Name _____

Date _____

Period _____

7. More Independent Practice

Solve the following equations. Show all work and circle your answers.

1. $16x + 39 = 3x$ **x = -3**

2. $9 + 3x = {}^-6x$ **x = -1**

3. $3x - 18 = 12x$ **x = -2**

4. $13m + 6 = {}^-4m$ **m = -$\frac{6}{17}$**

5. $3x = 77 - 8x$ **x = 7**

6. ${}^-5x = 27 + 4x$ **x = -3**

7. ${}^-11x = {}^-18 - 5x$ **x = 3**

8. ${}^-4c = 9 + 2c$ **c = -1$\frac{1}{2}$**

9. ${}^-16x = {}^-80 + 4x$ **x = 4**

10. $5w - 90 = 3w$ **w = 45**

11. $^-16x = 14x + 20$ $x = -\frac{2}{3}$ 12. $5y = ^-12y + 17$ $y = 1$

13. $^-7w = ^-8w - 99$ $w = ^-99$ 14. $8x = ^-8 - 6x$ $x = -\frac{4}{7}$

REVIEW

Solve the following problems. Show all work and circle your answers.

1. An aquarium measures 15 in by 24 in by 24 in. What is the volume of the aquarium? $V = lwh$

    8,640 in³

2. For the month of March, Mrs. Birch's average daily electrical cost was $0.87. How much was her bill for the entire month of March?

    $26.97

3. Simplify by combining like terms. $9x - 4x^3 + 6x^3 - 3x^3 - 2x^2y$

    $-x^3 - 2x^2y + 9x$

Algebra Ready

Try This 8

Word Problems

Write an equation for each of the following problems and solve it.

1. The sum of two numbers is 44. The second number is three times as large as the first. What are the two numbers?

   $x + 3x = 44$

   $x = 11$

   $3x = 33$

2. Diane has $16 more than Jessica. They have $96 together. How much does each girl have?

   $m + (m + 16) = 96$

   Jessica has $40. Diane has $56.

3. Phil's age is $\frac{5}{7}$ of 35. How old is Phil?

   $\frac{5}{7} \cdot 35 = x$

   Phil is 25.

4. The sum of two consecutive even numbers is 78. What are the numbers?

   $x + (x + 2) = 78$

   The numbers are 38 and 40.

5. The sum of three consecutive numbers is 48. What are the numbers?

   $x + (x + 1) + (x + 2) = 48$

   The numbers are 15, 16, and 17.

Write an equation for each of the following problems and solve it.

6. There are 200 students at 24th Street School. There are 22 more boys than girls. How many girls and boys are there in this school?

   $B = G + 22$

   $G + (G + 22) = 200$

   There are 89 girls and 111 boys.

7. At the football game, the Redwings gained 5 yards, then were pushed back 8 yards. They then moved ahead 42 yards. How far are they from where they started?

   $x = 5 - 8 + 42$

   $x = 39$

8. A rectangle is $\frac{1}{4}$ as wide as it is long. It is 20 feet long. What is the perimeter?

   $P = 2l \times 2w$

   $P = 2 \cdot 20 + 2 \cdot 20 \cdot \frac{1}{4}$

   $P = 50$ feet

9. Mr. Andrews has to make 500 copies of a 10-page document. There are two copiers in his office. One makes 5 copies a minute, and the other makes 20 copies a minute. How many hours will it take him to make the copies?

   $60(5 + 20)x = 500 \cdot 10$

   $3\frac{1}{3}$ hours (3 hours, 20 minutes)

10. Sammy has $2.25 in dimes and quarters. He has twice as many dimes as quarters. How many dimes and quarters does he have?

    $0.25 \cdot x + 0.10 \cdot 2x = \$2.25$

    Sammy has 5 quarters and 10 dimes.

Algebra Ready

Resources

Algebra Ready

Explanation: Order of Operations

Order of Operations

The order of operations gives you the rules for doing a math problem that requires more than one operation. It is the foundation for simplifying algebraic expressions and solving equations. If you do not follow the order of operations, the result could be a wrong answer.

Please **E**xcuse **M**y **D**ear **A**unt **S**ally is a mnemonic device to remember the order of operations.

① **P**lease stands for **P**arentheses (and other grouping symbols): ( ), [ ], { }.

② **E**xcuse stands for **E**xponents: $3^2, 5^2, 6^3, 2^2$, etc.

③ **M**y stands for **M**ultiplication.   (Done in order they appear from
   Dear stands for **D**ivision.   left to right in the problem)

④ **A**unt stands for **A**ddition.   (Done in order they appear from
   Sally stands for **S**ubtraction. left to right in the problem)

We first look at an expression and use the order of operations to determine what we do first, second, third, etc., to solve the equation.

Example 1:

First, look at the problem. Ask yourself what to do first, second, etc.

$17 - 12 \div 2 = ?$ Follow the order of operations. There are no ①parentheses and no ②exponents. Do ③multiplication or division from left to right. In this case, divide. Bring the other number down.

$17 - 6 = ?$ Next, ④add or subtract from left to right. In this case, subtract.

11 The answer is 11.

If you did not know the order of operations and subtracted first, you would have gotten the wrong answer.

$17 - 12 \div 2 = ?$ According to the order of operations, you divide first, but here we have incorrectly subtracted first.

$5 \div 2 = ?$

$2\frac{1}{2}$ The answer is wrong.

Example 2:

$2 \times 6^2 - 40 - 2 \times 4 = ?$ First, look at the problem. Ask yourself what to do first, second, etc.

Follow the order of operations. There are no ❶parentheses, so solve the ❷exponent first. Bring all other numbers down.

$2 \times 36 - 40 - 2 \times 4 = ?$ Do ❸multiplication or division from left to right. In this case, do both multiplications. Bring the other number down.

$72 - 40 - 8 = ?$ ❹Add or subtract from left to right. In this case, subtract.

$32 - 8 = ?$

24 The answer is 24.

Example 3:

$(2 \times 4 + 2) \times 2^2 \div (10 - 2) - 2 = ?$ First, look at the problem. Ask yourself what to do first, second, etc.

Follow the order of operations. Work within the ❶parentheses first. Always follow the same order of operations within parentheses. In the first set of parentheses, multiply…

$(8 + 2) \times 2^2 \div (10 - 2) - 2 = ?$ …then add.

$10 \times 2^2 \div (10 - 2) - 2 = ?$ Do the subtraction in the other set of parentheses.

$10 \times 2^2 \div 8 - 2 = ?$ ❷Exponents are next. Bring all other numbers down.

$10 \times 4 \div 8 - 2 = ?$ Do ❸multiplication or division from left to right. In this case, multiply first…

$40 \div 8 - 2 = ?$ …then divide.

$5 - 2 = ?$ ❹Add or subtract from left to right. In this case, subtract.

3 The answer is 3.

Vocabulary Words by Lesson

FLASHCARDS — PART 1

Lesson 1
sum
difference
product
quotient
order of operations

Lesson 2
exponent
base
power
estimate (verb)

Lesson 3
squared
cubed
percent
expression
grouping symbols

Lesson 4
simplify
order of operations
variable
substitute (verb)
evaluate

Lesson 5
equation
formula
area
diameter
radius
circumference
pi (or π)

Lesson 6
whole numbers
integers
opposites
zero (0)
number line

Lesson 7
volume
base
centimeter
plane

Lesson 8
sequence
negative number
digit
hundreds place
tens place

Lesson 9
range
data
addend
equivalent
order of operations

Lesson 10
consecutive numbers
sum
product
difference
quotient

Lesson 11
numerator
denominator
equation

Lesson 12
variable
inverse operations
substitute (verb)
algebra
isolate

Lesson 13
least common denominator (LCD)
improper fraction
inequality
mixed number
coefficient

Note: Italicized words appear in the lesson.

FLASHCARDS — PART 2

Lesson 14
triple
factor (noun)
greatest common factor (GCF)
unlike terms

Lesson 15
double
thousands place
average
mean

Lesson 16
term
coefficient
like terms
simplify
polynomial
binomial

Lesson 17
monomial
trinomial
symbol

Lesson 18
percent
divisible
volume
variable term

Lesson 19
least common denominator (LCD)
multiple
ratio
fraction
equivalent
real number

Lesson 20
meter
metric system
kilo-
tenths place
unit

Lesson 21
geometry
base
circumference
diameter
radius
trapezoid
parallelogram
evaluate

Lesson 22
factor (noun)
prime number
composite number
divisible

Lesson 23
inequality
equation
symbol

Lesson 24
least common denominator (LCD)
fraction
hundredths place
improper fraction
mixed number

Lesson 25
coefficient
decimal
inverse operations
value
like terms
real number

Lesson 26 (Review Words)
_(no cards)
fraction
quotient
product
geometry
decimal
sum
difference

Lesson 27
perimeter
area
circumference
volume

Lesson 28
monomial
distribute
coefficient
simplify
unlike terms
algebra

Lesson 29 (Review Words)
(no cards)

polynomial
variable
squared
exponent
monomial
estimate

Lesson 30 (Review Words)
(no cards)

grouping symbols
integer
composite number
factor
prime number

Lesson 31 (Review Words)
(no cards)

area
trinomial
like terms
pi
perimeter
binomial
unlike terms

Lesson 32 (Review Words)
(no cards)

equivalent
expression
denominator
integer
numerator

Lesson 33 (Review Words)
(no cards)

formula
ratio
distribute
coefficient
grouping symbols
cubed

Lesson 34 (Review Words)
(no cards)

unlike terms
evaluate
equation
binomial
inequality
value

Lesson 35 (Review Words)
(no cards)

coefficient
order of operations
unlike terms
distribute
polynomial

Lesson 36 (Review Words)
(no cards)

algebra
equation
difference
sequence
variable

Algebra Ready — Diagnostic Assessment—Subtest 1

Name: _____ **Grade:** _____ **Date:** _____ **Period:** _____

Pretest ☐   Posttest ☐

Vocabulary

Directions: Select a word from the list of words below to make each sentence complete. Put the correct answer in the column to the right. ☞

expression   area   exponent   plane   equation
data   integers   negative   sum
order of operations

1. The answer to an addition problem is called the _____.

2. The small raised number in 7^2 is called an _____.

3. A mathematical statement that stands for a given value is called an _____.

4. The _____ gives the rules for doing a problem that requires more than one operation.

5. The _____ is the space inside a closed, two-dimensional figure.

6. Counting numbers, their opposites, and 0 (…$^-2, ^-1, 0, ^+1, ^+2$…) are called _____.

7. A _____ is a flat surface, with no thickness, that extends forever.

8. A _____ number is any number found to the left of the 0 on a number line.

9. The numerical information that can be put into charts, tables, etc., for comparison is called _____.

Scoring: Subtests 1, 2, 3, 4 — 1 point is given for each correct subtest response for a possible score of 25. Total Score — Each subtest has a possible score of 25. All subtests are combined for a possible score of 100.

Subtest 1 _____ Subtest 2 _____ Subtest 3 _____ Subtest 4 _____

Total Score (all subtests): _____

Lessons	Answers
(1)	1. sum
(2)	2. exponent
(3)	3. expression
(4)	4. order of operations
(5)	5. area
(6)	6. integers
(7)	7. plane
(8)	8. negative
(9)	9. data
(1)	10. 1
(1)	11. 2
(2)	12. 50
(2)	13. 23
(3)	14. 9
(3)	15. 33
(4)	16. 6
(4)	17. 57
(5)	18. 12.56 in²
(5)	19. 31,400 cm²
(6)	20. ⁻8
(6)	21. ⁻6
(7)	22. 60
(8)	23. ⁻5
(9)	24. 5
(9)	25. 144

Total Correct: _____

Problem Solving

Find the value of the following expressions. Put answers in the answer column to the right on page 1.

10. $7 - 5 + 2 - 3$

11. $2 \times 3 \times 4/2/3/2$

12. $6^2 + 12 \div 2 \times 3 - 8 + 4$

13. $8 \times 4 - 4^2 + 14/2$

14. $[(6 \times 3) \div 3^2 - 1] + 2 \times 4$

15. $3(2 + 3^2)$

16. $25 \div 5^2 + (2 + 1)^2 - 4$

17. $4[(4 - 2) + 4^2] - 15$

Algebra Ready

Diagnostic Assessment—Subtest 1

Name: _____ **Grade:** _____ **Date:** _____ **Period:** _____

Solve the following problems ($\pi = 3.14$). Put answers in the answer column to the right on page 1.

18. Find the area of a circle with a radius of 2 in. Use the formula $A = \pi r^2$.

19. Find the area of a circle with a radius of 100 cm. Use the formula $A = \pi r^2$.

Find the values of the following expressions. For problems 20 and 21, use your *Algebra Ready* bookmark number line.

20. $^-3 - 5$

21. $^-8 + 7 - 6 + 5 - 4$

22. $(^-3)(5)(^-2)(2)$

23. $7(2 - 4) + 3^2$

24. $^-3 - (^-7) + 4 - 5 - (^-2)$

25. $(2 + 2)^2(5 + 4)$

Algebra Ready — Diagnostic Assessment—Subtest 2

Name: _____ Grade: _____ Date: _____ Period: _____

Pretest ☐   Posttest ☐

Vocabulary

Directions: Select a word from the list of words below to make each sentence complete. Put the correct answer in the column to the right. ☞

average   volume   algebra   product   equation
inequality   percent   coefficient   binomial
least common denominator (LCD)   simplify
greatest common factor (GCF)   inverse operation

1. The _____ is the answer to a multiplication problem.

2. A mathematical sentence that has two equivalent expressions separated by an equal sign is called an _____.

3. The _____ of addition is subtraction.

4. _____ is a part of mathematics that deals with variables, symbols, and numbers.

5. The _____ is the smallest multiple that two or more denominators have in common.

6. An _____ is a mathematical sentence in which <, >, ≤, or ≥ is used to compare two unequal values.

7. The _____ is the number directly in front of a variable. It is the number by which the variable is multiplied.

8. The _____ is the largest factor common to two or more numbers.

Lessons	Answers
(10)	1. *product*
(11)	2. *equation*
(12)	3. *inverse operation*
(12)	4. *algebra*
(13)	5. *LCD*
(13)	6. *inequality*
(13)	7. *coefficient*
(14)	8. *GCF*
(15)	9. *average*
(16)	10. *binomial*
(18)	11. *percent*
(18)	12. *volume*
(10)	13. *68*
(11)	14. *9*
(12)	15. *m = 7*
(13)	16. *m = 24*
(14)	17. *x = ⁻7*
(15)	18. *x = ⁻2*
(15)	19. *x = ⁻6*
(16)	20. *5h − 22y*
(16)	21. *20w − 7x + 7y*
(17)	22. *$5h^2 + a − 22y$*
(17)	23. *⁻12x + 12y*
(18)	24. *x = 3*
(18)	25. *m = 2*

Scoring: Subtests 1, 2, 3, 4 — 1 point is given for each correct subtest response for a possible score of 25.
Total Score — Each subtest has a possible score of 25. All subtests are combined for a possible score of 100.

Subtest 1 _____   Subtest 2 _____   Subtest 3 _____   Subtest 4 _____

Total Score (all subtests): _____

Total Correct: _____

9. An _____ is the sum of the values of a set of items divided by the number of items added.

10. A _____ is a polynomial with exactly two unlike terms.

11. A way to express a part of 100 is called a _____.

12. The amount of space inside a three-dimensional object is called _____.

Problem Solving

Find the value of the following expressions. Put answers in the answer column to the right on page 1.

13. $4[5 + 3(2 - 4)^2]$

14. $\dfrac{63 - 18}{-3 + 1 + 7}$

Find the value of the variable in the following equations.

15. $m - 24 = {-}17$

16. $-2m = {-}48$

17. $-3x + 6 = 27$

18. $7 = {-}5 - 6x$

Algebra Ready

Diagnostic Assessment—Subtest 2

Name: _____ Grade: _____ Date: _____ Period: _____

Solve or simplify the following problems. Put answers in the answer column to the right on page 1.

19. $^-4x + 3 = 27$

20. $^-7h - 12y - 10y + 12h$

21. $7y + 12w - 7x + 8w$

22. $^-7h^2 - 12y - 10y + 12h^2 + a$

23. $12y + 12w - 12x - 12w$

24. $^-4x + 33 = 7x$

25. $^-13m = {^-14} - 6m$

Algebra Ready — Diagnostic Assessment—Subtest 3

Name: _____ Grade: _____ Date: _____ Period: _____

Pretest ☐   Posttest ☐

Vocabulary

Directions: Select a word from the list below to make each sentence complete. Put the correct answer in the column to the right. ☞

inverse operations	inequality	meter	equation
3.14	coefficient	4.13	geometry
real number	equivalent	prime number	

1. Terms or expressions that have equal values are _____.

2. Any integer, decimal, or fraction is a _____.

3. A _____ is the basic unit of length in the metric system.

4. The study of points, lines, angles, planes, and shapes, and their relationships is called _____.

5. Pi equals _____.

6. A _____ is a positive whole number with only two factors: 1 and itself.

7. An _____ is a mathematical sentence in which $<, >, \leq$, or $\geq$ is used to compare two unequal values.

8. An _____ is a mathematical sentence that has two equivalent expressions separated by an equal sign.

9. The _____ is the number in front of a variable. It tells you how many of the variable you have.

10. _____ are operations that undo each other.

Scoring: Subtests 1, 2, 3, 4 — 1 point is given for each correct subtest response for a possible score of 25.
Total Score — Each subtest has a possible score of 25. All subtests are combined for a possible score of 100.

Subtest 1 _____ Subtest 2 _____ Subtest 3 _____ Subtest 4 _____

Total Score (all subtests): _____

Lessons	#	Answers
(19)	1.	equivalent
(19)	2.	real number
(20)	3.	meter
(21)	4.	geometry
(21)	5.	3.14
(22)	6.	prime number
(23)	7.	inequality
(23)	8.	equation
(25)	9.	coefficient
(25)	10.	inverse operations
(19)	11.	$x = {-4}$
(19)	12.	$n = {-2}$
(20)	13.	no solution
(20)	14.	k = all real numbers
(21)	15.	$A = 128\ m^2$
(21)	16.	$C = 31.4\ in$
(22)	17.	$m = 1\frac{1}{2}$
(22)	18.	$k = 2$
(23)	19.	$x < 7$
(24)	20.	$a = 27$
(25)	21.	$n = 2$
(26)	22.	$x = 3$
(26)	23.	$d = 7.5$
(27)	24.	$A = 48\ m^2$
(27)	25.	$V = 512\ cm^3$

Total Correct: _____

Problem Solving

Solve the following problems. Put answers in the answer column to the right on page 1.

11. $-4x + 3 = 27 + 2x$

12. $-6n - 14 = 18 + 10n$

13. $12 - 2m = -2m - 18$

14. $7k + 15 = 15 + 7k$

15. Find the area.

    $A = \dfrac{h(B + b)}{2}$

    b = 12 m
    h = 8 m
    B = 20 m

16. Find the circumference.

    $C = 2\pi r$

    r = 5 in

·152·

Algebra Ready

Diagnostic Assessment—Subtest 3

Name: _____ Grade: _____ Date: _____ Period: _____

Solve the following problems. Put answers in the answer column to the right on page 1.

17. $^-12 - 2m = 2m - 18$

18. $^-7k + 15 = {^-13} + 7k$

19. $7 - 6x > {^-35}$

20. $\frac{2}{3}a + 9 = \frac{1}{3}a + 18$

21. $1.2n - 3.44 = {^-0.8n} + 0.56$

22. $\frac{2}{3}x - \frac{1}{2} = \frac{1}{6}x + 1$

Solve the following problems. Put answers in the answer column to the right on page 1.

23. $0.4 + 0.2d = 1.9$

24. $A = \frac{1}{2}bh$ $h = 8\ m$ $b = 12\ m$

25. $V = s^3$ $s = 8\ cm$

Algebra Ready
Diagnostic Assessment—Subtest 4

Name: _____ Grade: _____ Date: _____ Period: _____

Pretest ☐   Posttest ☐

Vocabulary

Directions: Select a word from the list of words below to make each sentence complete. Put the correct answer in the column to the right. ☞

simplify	value	prime number
exponent	monomial	ratio
distribute		

1. A _____ is an algebraic expression with exactly one term.

2. To _____ is to combine like terms and put the answer in lowest terms.

3. A _____ is a way of comparing numbers: $\frac{3}{4}$, 3:4, 3 to 4.

4. To _____ is to multiply each term within a set of grouping symbols by the term immediately preceding or following that grouping symbol.

5. A _____ is an assigned amount.

Problem Solving

Simplify the following problems.

6. $^-3x \cdot 3$ 7. $(23b)(^-3)$

_____

_____

_____

_____

Scoring: Subtests 1, 2, 3, 4 — 1 point is given for each correct subtest response for a possible score of 25.
Total Score — Each subtest has a possible score of 25. All subtests are combined for a possible score of 100.

Subtest 1 _____   Subtest 2 _____   Subtest 3 _____   Subtest 4 _____

Total Score (all subtests): _____

Lessons	Answers
(28)	1. _monomial_
(28)	2. _simplify_
(33)	3. _ratio_
(33)	4. _distribute_
(34)	5. _value_
(28)	6. ^-9x
(28)	7. ^-69b
(29)	8. $100n + 25$
(29)	9. $^-9v - 18$
(30)	10. $11c + 5$
(30)	11. $48v + 32$
(30)	12. $^-3x + 44$
(31)	13. a
(31)	14. $2n^3 + n^2 + n + 14$
(32)	15. $^-2g + 8h$
(32)	16. $^-6p^2$
(33)	17. $h = 3$
(33)	18. $3\frac{1}{3}$
(33)	19. $w = ^-1$
(34)	20. $x = 11$
(34)	21. $v = 0$
(35)	22. $g = ^-\frac{1}{2}$
(36)	23. $x = 23\frac{1}{8}$
(36)	24. 43
(36)	25. $43 \& 57$

Total Correct: _____

Simplify the following problems. Put answers in the answer column to the right on page 1.

8. $5(20n + 5)$

9. $^-3(3v + 4 + 2)$

10. $2(4c - 2) + 3(c + 3)$

11. $7(4v + 6) - 5(2 - 4v)$

12. $7(8 - 3x) + 3(8x - 2x - 4)$

13. $(^-12h^2 - 12y) + (12y + 12h^2 + a)$

14. $(n^2 + 2n^3 + 7) + (7 + n)$

15. $(3h + 4g) - (6g - 5h)$

Algebra Ready

Diagnostic Assessment—Subtest 4

Name: _____ Grade: _____ Date: _____ Period: _____

Solve or simplify the following problems. Put answers in the answer column to the right on page 1.

16. $(24p^2 - 16t - 15) - (30p^2 - 16t - 15)$

17. $3(6h + 2) - 5 = 55$

18. $^-7(^-3x + 8) = 14$

19. $5(-w - 3) = {^-}10$

20. $11 = 7x - 66$

21. $4(v + 9) = 6(2v + 2 + 8 \div 2)$

Solve the following problems. Put answers in the answer column to the right on page 1.

22. $6(2g - 4) = 6(3g - 5) + 2(3g + 6)$

23. When the stock market opened today, shares of Apex Oil were valued at $26\frac{3}{8}$ per share. When the market closed, the shares had dropped $3\frac{1}{4}$ points. What was the new value?

24. Benita has 12 more than twice as many customers for her beauty shop as she had before she started advertising. She now has 98 customers. How many customers did she have when she started?

25. Separate 100 into two parts so that the larger number is 14 more than the smaller number. What are the numbers?

Algebra Ready

Diagnostic Assessment Report—Subtest 1

Teacher: **Mr. Smyth** Class: **Math** # Students: **23** Period: **8** Date: **12/3/02**

Pretest Sample

Student	Vocabulary Lessons										Application Lessons										Score	Remarks
	1	2	3	4	5	6	7	8	9	1	2	3	4	5	6	7	8	9		
Garcia	◆	◆	◆	◆	◆	◆	◆	◆	◆					◆	◆	◆		◆	13	
Johnson	◆	◆	◆	◆	◆	◆	◆	◆	◆	◆						◆	◆		14	
Williams	◆	◆		◆	◆	◆	◆	◆	◆	◆		◆			◆				11	
Fernandez	◆	◆	◆	◆	◆	◆	◆	◆	◆	◆			◆						10	
Peña	◆	◆	◆	◆	◆	◆	◆	◆	◆										9	
Miller	◆	◆		◆	◆	◆	◆		◆										7	
Hall	◆				◆	◆	◆	◆	◆										7	
Meadows		◆	◆	◆		◆		◆	◆	◆									7	
Carter	◆								◆			◆	◆		◆				7	
Montoya	◆	◆																	5	
Hayashi					◆	◆	◆	◆											5	
Andersson	◆						◆		◆	◆									5	
Chong	◆					◆			◆	◆					◆				5	
Thomson	◆	◆				◆			◆	◆					◆				5	
Nguyen	◆																		4	
Hammill									◆										4	
Feinstein	◆	◆												◆					3	
Rogers									◆										2	
O'Brien	◆								◆										2	
McNeil			◆																2	
TOTAL																				

Algebra Ready
Diagnostic Assessment Report—Subtest 1

Teacher: **Mr. Smyth** Class: **Math** # Students: **23** Period: **8** Date: **12/3/02**

Student	Vocabulary Lessons									Application Lessons									Score	Remarks
	1	2	3	4	5	6	7	8	9	1	2	3	4	5	6	7	8	9	9	
Schmitt	♦																		2	
Castilla	♦				♦														2	
Jacobson	♦									♦									1	
TOTAL	18	10	5	9	11	10	12	11	13	13	1			2	1	1		4	3	2	1

·160·

Algebra Ready

Diagnostic Assessment Report—Subtest 1

Teacher: _____ Class: _____ # Students: _____ Period: _____ Date: _____

Student	Vocabulary Lessons									Application Lessons									Score	Remarks
	1	2	3	4	5	6	7	8	9	1	2	3	4	5	6	7	8	9		
TOTAL																				

Algebra Ready

Diagnostic Assessment Report—Subtest 2

Teacher: _____ Class: _____ # Students: _____ Period: _____ Date: _____

Student	Vocabulary Lessons										Application Lessons										Score	Remarks
	10	11	12	12	13	13	13	14	15	16	18	18	10	11	12	13	14	15	15	16	16	17	17	18	18		

TOTAL

Algebra Ready

Diagnostic Assessment Report—Subtest 3

Teacher: _____ Class: _____ # Students: _____ Period: _____ Date: _____

Student	Vocabulary Lessons							Application Lessons									Score	Remarks
	19	20	21	22	23	25		19	20	21	22	23	24	25	26	27		
TOTAL																		

Algebra Ready — Diagnostic Assessment Report—Subtest 4

Teacher: _____   Class: _____   # Students: _____   Period: _____   Date: _____

Student	Vocabulary Lessons					Application Lessons																		Score	Remarks
	28	28	33	33	34	28	28	29	29	30	30	30	31	31	32	32	33	33	33	34	34	35	36	36	36
TOTAL																									

Algebra Ready — Skill Practice 1   +

Name _____
Date _____
Period _____

Addition I

36	50	9	61
+1	+29	+4	+8
37	**79**	**13**	**69**
60	16	12	12
+3	+13	+0	+11
63	**29**	**12**	**23**
73	36	73	47
+20	+32	+121	+22
93	**68**	**194**	**69**
7	89	9	50
+22	+1	+7	+33
29	**90**	**16**	**83**
5	61	69	14
+1	+17	+10	+3
6	**78**	**79**	**17**
81	13	71	54
+13	+13	+5	+22
94	**26**	**76**	**76**
43	9	9	54
+43	+3	+9	+32
86	**12**	**18**	**86**

Time: _____ min.

Addition II

46	82	86	33
+23	+13	+10	+11
69	**95**	**96**	**44**
10	14	85	9
+11	+5	+14	+6
21	**19**	**99**	**15**
77	50	62	6
+22	+47	+16	+2
99	**97**	**78**	**8**
37	8	10	18
+10	+2	+11	+3
47	**10**	**21**	**21**
45	17	45	50
+1	+10	+17	+3
46	**27**	**62**	**53**
65	23	26	75
+13	+5	+8	+7
78	**28**	**34**	**82**
34	29	111	14
+43	+9	+98	+75
77	**38**	**209**	**89**

Time: _____ min.

Addition III

37	51	10	62
+2	+30	+5	+9
39	**81**	**15**	**71**
61	17	13	13
+4	+14	+1	+12
65	**31**	**14**	**25**
74	37	74	48
+21	+33	+122	+23
95	**70**	**196**	**71**
8	90	10	51
+23	+2	+8	+34
31	**92**	**18**	**85**
6	62	70	15
+2	+18	+11	+4
8	**80**	**81**	**19**
82	14	72	55
+14	+14	+6	+23
96	**28**	**78**	**78**
44	10	10	55
+44	+4	+10	+33
88	**14**	**20**	**88**

Time: _____ min.

Addition IV

47	83	87	34
+24	+14	+11	+10
71	**97**	**98**	**44**
11	15	86	10
+12	+6	+15	+7
23	**21**	**101**	**17**
78	51	63	7
+23	+48	+17	+3
101	**99**	**80**	**10**
38	9	11	19
+11	+3	+12	+4
49	**12**	**23**	**23**
46	18	46	51
+2	+11	+18	+4
48	**29**	**64**	**55**
66	24	27	76
+14	+6	+9	+8
80	**30**	**36**	**84**
35	30	112	15
+44	+10	+99	+76
79	**40**	**211**	**91**

Time: _____ min.

Algebra Ready — Skill Practice 2 +

Name _____
Date _____
Period _____

Addition I

31 +26 **57**	10 +10 **20**	40 +40 **80**	56 +26 **82**
22 +11 **33**	50 +69 **119**	43 +17 **60**	29 +62 **91**
48 +4 **52**	49 +11 **60**	18 +8 **26**	11 +19 **30**
33 +77 **110**	85 +99 **184**	63 +34 **97**	25 +48 **73**
18 +3 **21**	25 +25 **50**	87 +4 **91**	14 +66 **80**
37 +80 **117**	50 +45 **95**	73 +17 **90**	48 +26 **74**
41 +90 **131**	90 +40 **130**	39 +42 **81**	80 +60 **140**

Time: _____ min.

Addition II

57 +73 **130**	46 +7 **53**	29 +13 **42**	64 +18 **82**
44 +72 **116**	39 +84 **123**	85 +7 **92**	50 +91 **141**
17 +48 **65**	47 +13 **60**	27 +18 **45**	84 +61 **145**
68 +3 **71**	24 +27 **51**	17 +17 **34**	73 +8 **81**
19 +16 **35**	67 +13 **80**	28 +7 **35**	94 +17 **111**
47 +19 **66**	13 +31 **44**	41 +14 **55**	76 +8 **84**
35 +44 **79**	30 +10 **40**	112 +99 **211**	15 +76 **91**

Time: _____ min.

Addition III

47 +12 **59**	61 +40 **101**	20 +15 **35**	72 +19 **91**
71 +14 **85**	27 +24 **51**	23 +11 **34**	23 +22 **45**
84 +31 **115**	47 +43 **90**	84 +132 **216**	58 +33 **91**
18 +33 **51**	100 +12 **112**	20 +18 **38**	61 +44 **105**
16 +12 **28**	72 +28 **100**	80 +21 **101**	25 +14 **39**
92 +24 **116**	24 +24 **48**	82 +16 **98**	65 +33 **98**
54 +54 **108**	20 +14 **34**	20 +20 **40**	65 +43 **108**

Time: _____ min.

Addition IV

57 +34 **91**	93 +24 **117**	97 +21 **118**	44 +20 **64**
21 +22 **43**	25 +16 **41**	96 +25 **121**	20 +17 **37**
88 +33 **121**	61 +58 **119**	73 +27 **100**	17 +13 **30**
48 +21 **69**	19 +13 **32**	21 +22 **43**	29 +14 **43**
56 +12 **68**	28 +21 **49**	56 +28 **84**	61 +14 **75**
76 +24 **100**	34 +16 **50**	37 +19 **56**	86 +18 **104**
45 +54 **99**	40 +20 **60**	122 +109 **231**	25 +86 **111**

Time: _____ min.

Algebra Ready — Skill Practice 3   +

Name _____

Date _____

Period _____

Addition I

11 +13 **24**	20 +32 **52**	48 +53 **101**	32 +64 **96**
77 +60 **137**	55 +49 **104**	56 +78 **134**	93 +39 **132**
11 +99 **110**	74 +64 **138**	53 +34 **87**	90 +17 **107**
75 +25 **100**	33 +67 **100**	396 +404 **800**	569 +322 **891**
596 +232 **828**	667 +348 **1,015**	597 +668 **1,265**	345 +549 **894**
197 +204 **401**	567 +664 **1,231**	928 +113 **1,041**	925 +606 **1,531**
733 +248 **981**	533 +678 **1,211**	434 +547 **981**	478 +344 **822**

Time: _____ min.

Addition II

815 +597 **1,412**	856 +678 **1,534**	471 +777 **1,248**	567 +664 **1,231**
185 +369 **554**	550 +330 **880**	352 +352 **704**	120 +506 **626**
403 +403 **806**	774 +111 **885**	456 +156 **612**	729 +203 **932**
336 +639 **975**	104 +401 **505**	442 +304 **746**	180 +180 **360**
111 +222 **333**	250 +250 **500**	123 +819 **942**	569 +333 **902**
118 +115 **233**	654 +120 **774**	141 +214 **355**	376 +128 **504**
335 +144 **479**	230 +310 **540**	212 +399 **611**	115 +276 **391**

Time: _____ min.

Addition III

147	361	220	172
+212	+140	+315	+219
359	**501**	**535**	**391**

371	227	123	323
+114	+324	+211	+122
485	**551**	**334**	**445**

284	347	184	358
+331	+243	+232	+333
615	**590**	**416**	**691**

118	200	320	161
+233	+312	+218	+144
351	**512**	**538**	**305**

216	372	180	225
+312	+228	+121	+314
528	**600**	**301**	**539**

392	124	282	165
+224	+124	+316	+233
616	**248**	**598**	**398**

354	220	120	365
+354	+114	+220	+343
708	**334**	**340**	**708**

Time: _____ min.

Addition IV

157	293	397	144
+134	+324	+221	+220
291	**617**	**618**	**364**

321	225	196	320
+322	+116	+225	+317
643	**341**	**421**	**637**

488	261	173	317
+333	+158	+227	+413
821	**419**	**400**	**730**

348	119	221	429
+221	+413	+322	+114
569	**532**	**543**	**543**

356	228	456	161
+212	+321	+128	+414
568	**549**	**584**	**575**

276	334	437	586
+324	+216	+119	+218
600	**550**	**556**	**804**

345	740	222	925
+654	+320	+609	+186
999	**1,060**	**831**	**1,111**

Time: _____ min.

Algebra Ready — Skill Practice 4 +

Name _____

Date _____

Period _____

Addition I

555 +444 **999**	222 +555 **777**	336 +265 **601**	321 +213 **534**
336 +226 **562**	222 +559 **781**	357 +159 **516**	239 +456 **695**
205 +520 **725**	103 +103 **206**	554 +332 **886**	633 +200 **833**
781 +156 **937**	258 +369 **627**	148 +598 **746**	544 +455 **999**
336 +336 **672**	889 +120 **1,009**	654 +108 **762**	601 +319 **920**
21,234 +54,563 **75,797**	65,236 +56,521 **121,757**	11,023 +25,554 **36,577**	
41,254 +21,254 **62,508**	93,209 +19,632 **112,841**	23,005 +52,005 **75,010**	

Time: _____ min.

Addition II

3,503 +3,650 **7,153**	2,222 +3,658 **5,880**	3,502 +4,503 **8,005**	5,023 +6,325 **11,348**
8,920 +6,320 **15,240**	5,556 +2,228 **7,784**	5,688 +2,222 **7,910**	6,654 +3,365 **10,019**
5,555 +5,555 **11,110**	4,563 +4,563 **9,126**	9,635 +9,635 **19,270**	1,052 +1,052 **2,104**
5,600 +5,600 **11,200**	5,555 +9,632 **15,187**	6,663 +5,555 **12,218**	3,369 +6,602 **9,971**
1,147 +2,352 **3,499**	1,235 +6,677 **7,912**	5,688 +2,222 **7,910**	2,580 +2,580 **5,160**
78,945 +61,230 **140,175**	65,498 +12,369 **77,867**	12,141 +34,214 **46,355**	
56,335 +78,144 **134,479**	91,230 +23,310 **114,540**	45,212 +67,399 **112,611**	

Time: _____ min.

Addition III

656	323	437	422
+545	+656	+366	+314
1,201	**979**	**803**	**736**

437	323	458	330
+327	+650	+250	+557
764	**973**	**708**	**887**

306	204	655	734
+621	+204	+433	+301
927	**408**	**1,088**	**1,035**

882	359	249	645
+257	+460	+699	+556
1,139	**819**	**948**	**1,201**

437	980	755	702
+437	+221	+209	+410
874	**1,201**	**964**	**1,112**

31,235	75,237	21,024
+64,564	+66,522	+35,555
95,799	**141,759**	**56,579**

51,255	83,208	33,006
+31,255	+29,633	+62,006
82,510	**112,841**	**95,012**

Time: _____ min.

Addition IV

4,504	1,223	2,501	6,024
+4,651	+2,659	+3,502	+7,326
9,155	**3,882**	**6,003**	**13,350**

2,928	4,555	4,687	5,653
+8,322	+1,227	+4,223	+2,364
11,250	**5,782**	**8,910**	**8,017**

6,554	5,562	8,634	3,051
+4,556	+3,562	+1,634	+5,051
11,110	**9,124**	**10,268**	**8,102**

3,609	4,554	5,662	2,368
+7,609	+8,631	+4,554	+5,601
11,218	**13,185**	**10,216**	**7,969**

2,146	2,234	4,687	1,589
+1,351	+5,676	+1,221	+1,589
3,497	**7,910**	**5,908**	**3,178**

68,944	55,497	22,140
+51,239	+22,368	+24,213
120,183	**77,865**	**46,353**

46,334	81,239	35,211
+68,143	+13,319	+57,398
114,477	**94,558**	**92,609**

Time: _____ min.

Algebra Ready — Skill Practice 1 −

Name _____

Date _____

Period _____

Subtraction I

36 − 1 **35**	50 − 29 **21**	9 − 4 **5**	61 − 8 **53**
60 − 3 **57**	16 − 13 **3**	12 − 0 **12**	12 − 11 **1**
73 − 20 **53**	36 − 32 **4**	73 − 11 **62**	47 − 22 **25**
22 − 7 **15**	89 − 1 **88**	9 − 7 **2**	50 − 33 **17**
5 − 1 **4**	61 − 17 **44**	69 − 10 **59**	14 − 3 **11**
81 − 13 **68**	13 − 13 **0**	71 − 5 **66**	54 − 22 **32**
43 − 40 **3**	9 − 8 **1**	19 − 9 **10**	54 − 32 **22**

Time: _____ min.

Subtraction II

46 − 23 **23**	82 − 13 **69**	86 − 10 **76**	33 − 11 **22**
20 − 11 **9**	94 − 2 **92**	14 − 5 **9**	9 − 6 **3**
77 − 22 **55**	50 − 37 **13**	62 − 16 **46**	6 − 2 **4**
37 − 10 **27**	8 − 2 **6**	40 − 11 **29**	18 − 3 **15**
45 − 1 **44**	17 − 10 **7**	45 − 17 **28**	50 − 3 **47**
65 − 13 **52**	23 − 5 **18**	26 − 8 **18**	75 − 7 **68**
43 − 34 **9**	29 − 9 **20**	111 − 98 **13**	75 − 14 **61**

Time: _____ min.

Subtraction III

37	51	10	62
− 2	− 30	− 5	− 9
35	**21**	**5**	**53**

61	17	13	13
− 4	− 14	− 1	− 12
57	**3**	**12**	**1**

74	37	122	48
− 21	− 33	− 74	− 23
53	**4**	**48**	**25**

23	90	10	51
− 8	− 2	− 8	− 34
15	**88**	**2**	**17**

6	62	70	15
− 2	− 18	− 11	− 4
4	**44**	**59**	**11**

82	14	72	55
− 14	− 14	− 6	− 23
68	**0**	**66**	**32**

44	10	10	55
− 44	− 4	− 10	− 33
0	**6**	**0**	**22**

Time: _____ min.

Subtraction IV

47	83	87	34
− 24	− 14	− 11	− 10
23	**69**	**76**	**24**

12	15	86	10
− 11	− 6	− 15	− 7
1	**9**	**71**	**3**

78	51	63	7
− 23	− 48	− 17	− 3
55	**3**	**46**	**4**

38	9	21	19
− 11	− 3	− 12	− 4
27	**6**	**9**	**15**

46	18	46	51
− 2	− 11	− 18	− 4
44	**7**	**28**	**47**

66	24	27	76
− 14	− 6	− 9	− 8
52	**18**	**18**	**68**

65	30	112	115
− 44	− 10	− 99	− 76
21	**20**	**13**	**39**

Time: _____ min.

Algebra Ready — Skill Practice 2

Name _____
Date _____
Period _____

Subtraction I

31	100	40	56
−26	−10	−24	−26
5	**90**	**16**	**30**

22	80	43	79
−11	−69	−17	−62
11	**11**	**26**	**17**

48	49	18	31
−4	−11	−8	−19
44	**38**	**10**	**12**

133	85	63	85
−77	−59	−34	−48
56	**26**	**29**	**37**

18	25	87	64
−3	−15	−4	−46
15	**10**	**83**	**18**

97	50	73	48
−60	−45	−17	−26
37	**5**	**56**	**22**

41	90	89	80
−9	−40	−42	−60
32	**50**	**47**	**20**

Time: _____ min.

Subtraction II

57	46	29	64
−57	−7	−13	−18
0	**39**	**16**	**46**

144	39	85	50
−72	−14	−7	−21
72	**25**	**78**	**29**

97	47	27	84
−48	−13	−18	−61
49	**34**	**9**	**23**

68	74	87	73
−3	−27	−37	−8
65	**47**	**50**	**65**

19	67	28	94
−16	−13	−7	−17
3	**54**	**21**	**77**

47	53	41	76
−19	−31	−14	−8
28	**22**	**27**	**68**

55	30	112	115
−44	−10	−99	−76
11	**20**	**13**	**39**

Time: _____ min.

Subtraction III

47 − 12 **35**	61 − 40 **21**	20 − 15 **5**	72 − 19 **53**
71 − 14 **57**	27 − 24 **3**	23 − 11 **12**	23 − 22 **1**
84 − 31 **53**	47 − 43 **4**	184 − 32 **152**	58 − 33 **25**
78 − 33 **45**	100 − 12 **88**	20 − 18 **2**	61 − 44 **17**
16 − 12 **4**	72 − 28 **44**	80 − 21 **59**	25 − 14 **11**
92 − 24 **68**	24 − 24 **0**	82 − 16 **66**	65 − 33 **32**
84 − 54 **30**	20 − 14 **6**	90 − 20 **70**	65 − 43 **22**

Time: _____ min.

Subtraction IV

57 − 34 **23**	93 − 24 **69**	97 − 21 **76**	44 − 20 **24**
71 − 22 **49**	25 − 16 **9**	96 − 25 **71**	20 − 17 **3**
88 − 33 **55**	61 − 58 **3**	73 − 27 **46**	17 − 13 **4**
48 − 21 **27**	19 − 13 **6**	81 − 22 **59**	29 − 14 **15**
56 − 12 **44**	28 − 21 **7**	56 − 28 **28**	61 − 14 **47**
76 − 24 **52**	34 − 16 **18**	37 − 19 **18**	86 − 18 **68**
95 − 54 **41**	40 − 20 **20**	122 − 109 **13**	125 − 86 **39**

Time: _____ min.

Algebra Ready — Skill Practice 3 −

Name _____
Date _____
Period _____

Subtraction I

367	732	753	864
− 155	− 220	− 448	− 732
212	**512**	**305**	**132**
277	555	878	493
− 260	− 349	− 556	− 439
17	**206**	**322**	**54**
799	474	753	390
− 111	− 264	− 234	− 17
688	**210**	**519**	**373**
675	567	404	569
− 225	− 133	− 396	− 322
450	**434**	**8**	**247**
596	667	597	345
− 232	− 348	− 468	− 249
364	**319**	**129**	**96**
397	867	928	925
− 204	− 664	− 113	− 606
193	**203**	**815**	**319**
733	533	734	478
− 248	− 278	− 547	− 344
485	**255**	**187**	**134**

Time: _____ min.

Subtraction II

815	856	471	967
− 597	− 678	− 277	− 664
218	**178**	**194**	**303**
185	550	552	820
− 69	− 330	− 352	− 506
116	**220**	**200**	**314**
403	774	456	729
− 403	− 111	− 156	− 203
0	**663**	**300**	**526**
336	604	442	180
− 39	− 401	− 304	− 23
297	**203**	**138**	**157**
433	850	819	569
− 222	− 180	− 123	− 333
211	**670**	**696**	**236**
118	654	214	376
− 115	− 120	− 141	− 128
3	**534**	**73**	**248**
335	310	399	276
− 144	− 230	− 212	− 115
191	**80**	**187**	**161**

Time: _____ min.

Subtraction III

212	361	315	219
− 147	− 140	− 220	− 172
65	**221**	**95**	**47**
371	324	211	323
− 114	− 247	− 123	− 122
257	**77**	**88**	**201**
331	347	232	358
− 284	− 243	− 184	− 333
47	**104**	**48**	**25**
233	312	320	161
− 118	− 200	− 218	− 144
115	**112**	**102**	**17**
312	372	180	314
− 216	− 228	− 121	− 225
96	**144**	**59**	**89**
392	124	316	233
− 224	− 124	− 282	− 165
168	**0**	**34**	**68**
354	220	220	365
− 354	− 114	− 120	− 343
0	**106**	**100**	**22**

Time: _____ min.

Subtraction IV

157	324	397	220
− 134	− 293	− 221	− 144
23	**31**	**176**	**76**
322	225	225	320
− 321	− 116	− 196	− 317
1	**109**	**29**	**3**
488	261	227	413
− 333	− 158	− 173	− 317
155	**103**	**54**	**96**
348	413	322	429
− 221	− 119	− 221	− 114
127	**294**	**101**	**315**
356	321	456	414
− 212	− 228	− 128	− 161
144	**93**	**328**	**253**
324	334	437	586
− 276	− 216	− 119	− 218
48	**118**	**318**	**368**
654	740	609	925
− 325	− 320	− 222	− 186
329	**420**	**387**	**739**

Time: _____ min.

Algebra Ready   Skill Practice 4   −

Name _____

Date _____

Period _____

Subtraction I

555	722	336	321
− 444	− 555	− 265	− 213
111	**167**	**71**	**108**

336	622	357	839
− 226	− 559	− 159	− 456
110	**63**	**198**	**383**

205	456	554	633
− 120	− 103	− 332	− 200
85	**353**	**222**	**433**

781	658	148	544
− 156	− 369	− 98	− 455
625	**289**	**50**	**89**

446	889	654	601
− 336	− 120	− 108	− 319
110	**769**	**546**	**282**

71,234	65,236	31,023
− 54,563	− 56,521	− 25,554
16,671	**8,715**	**5,469**

41,254	93,209	73,005
− 21,254	− 19,632	− 52,005
20,000	**73,577**	**21,000**

Time: _____ min.

Subtraction II

3,803	2,222	3,502	7,023
− 3,650	− 658	− 2,103	− 6,325
153	**1,564**	**1,399**	**698**

8,920	5,556	5,688	6,654
− 6,320	− 2,228	− 2,222	− 3,365
2,600	**3,328**	**3,466**	**3,289**

5,555	5,563	9,000	3,052
− 555	− 4,563	− 7,635	− 2,552
5,000	**1,000**	**1,365**	**500**

5,600	9,632	6,663	8,369
− 5,600	− 5,555	− 5,555	− 6,602
0	**4,077**	**1,108**	**1,767**

1,147	6,735	5,688	2,580
− 352	− 6,677	− 2,222	− 1,443
795	**58**	**3,466**	**1,137**

78,945	65,498	42,141
− 61,230	− 12,369	− 34,214
17,715	**53,129**	**7,927**

96,335	91,230	85,212
− 78,144	− 23,310	− 67,399
18,191	**67,920**	**17,813**

Time: _____ min.

Subtraction III

654	621	235	220
− 343	− 454	− 164	− 112
311	**167**	**71**	**108**

235	521	256	738
− 125	− 450	− 58	− 355
110	**71**	**198**	**383**

104	355	453	532
− 29	− 202	− 231	− 109
75	**153**	**222**	**423**

680	557	247	443
− 255	− 268	− 197	− 354
425	**289**	**50**	**89**

345	788	553	500
− 235	− 229	− 207	− 218
110	**559**	**346**	**282**

61,233	55,235	41,022
− 44,562	− 46,520	− 35,553
16,671	**8,715**	**5,469**

31,253	83,208	63,004
− 11,253	− 29,631	− 42,004
20,000	**53,577**	**21,000**

Time: _____ min.

Subtraction IV

2,802	1,221	2,501	6,022
− 2,659	− 557	− 1,102	− 5,324
143	**664**	**1,399**	**698**

7,929	4,555	4,687	5,655
− 5,329	− 1,227	− 1,221	− 2,364
2,600	**3,328**	**3,466**	**3,291**

4,554	4,562	8,009	2,051
− 454	− 3,562	− 6,634	− 1,551
4,100	**1,000**	**1,375**	**500**

4,609	8,631	5,662	7,368
− 4,609	− 4,554	− 4,554	− 5,601
0	**4,077**	**1,108**	**1,767**

2,146	5,734	4,687	3,589
− 251	− 5,676	− 1,221	− 2,442
1,895	**58**	**3,466**	**1,147**

68,944	55,497	52,142
− 51,239	− 22,368	− 44,213
17,705	**33,129**	**7,929**

86,334	81,231	75,211
− 68,143	− 13,319	− 57,390
18,191	**67,912**	**17,821**

Time: _____ min.

Algebra Ready — Skill Practice 1 +/−

Name _____

Date _____

Period _____

Addition/Subtraction I

106 + 304 **410**	123 + 99 **222**	324 − 106 **218**	809 + 100 **909**
322 − 111 **211**	237 + 692 **929**	212 + 464 **676**	869 − 125 **744**
625 − 100 **525**	665 + 334 **999**	525 + 125 **650**	533 − 100 **433**
4,117 + 2,106 **6,223**	3,396 − 2,196 **1,200**	6,904 − 2,431 **4,473**	1,544 + 9,455 **10,999**
2,446 + 3,336 **5,782**	9,889 − 7,120 **2,769**	9,654 − 5,108 **4,546**	2,601 + 5,319 **7,920**
12,164 + 13,421 **25,585**	62,371 − 25,257 **37,114**	72,890 − 60,051 **12,839**	
23,251 + 61,003 **84,254**	17,971 + 61,472 **79,443**	32,143 + 66,908 **99,051**	

Time: _____ min.

Addition/Subtraction II

342 + 600 **942**	625 − 525 **100**	626 − 232 **394**	201 + 642 **843**
400 − 100 **300**	649 − 102 **547**	141 +666 **807**	654 + 365 **1,019**
2,111 + 6,842 **8,953**	3,204 − 2,156 **1,048**	9,000 + 1,369 **10,369**	8,052 − 6,554 **1,498**
5,600 − 5,600 **0**	9,632 − 5,555 **4,077**	6,663 − 5,555 **1,108**	8,369 − 6,602 **1,767**
80,701 − 79,701 **1,000**	42,580 + 42,580 **85,160**	96,263 + 11,298 **107,561**	
78,845 + 61,230 **140,075**	64,115 − 11,115 **53,000**	86,868 + 86,868 **173,736**	
93,670 − 32,960 **60,710**	65,498 + 12,369 **77,867**	69,699 − 32,111 **37,588**	

Time: _____ min.

Addition/Subtraction III

207	224	223	900
+ 405	+ 190	− 15	+ 201
612	**414**	**208**	**1,101**

221	338	313	768
− 10	+ 793	+ 565	− 24
211	**1,131**	**878**	**744**

524	564	424	432
− 98	+ 233	+ 227	− 102
426	**797**	**651**	**330**

5,118	2,395	5,903	2,545
+ 3,107	− 1,195	− 1,430	+ 8,456
8,225	**1,200**	**4,473**	**11,001**

3,447	8,888	8,653	1,600
+ 4,337	− 6,129	− 4,107	+ 6,318
7,784	**2,759**	**4,546**	**7,918**

22,165	52,370	62,899
+ 23,422	− 15,256	− 50,050
45,587	**37,114**	**12,849**

33,252	27,972	42,144
+ 71,004	+ 71,473	+ 76,909
104,256	**99,445**	**119,053**

Time: _____ min.

Addition/Subtraction IV

443	524	525	302
+ 701	− 424	− 131	+ 743
1,144	**100**	**394**	**1,045**

309	548	242	755
− 209	− 203	+767	+ 466
100	**345**	**1,009**	**1,221**

3,112	4,203	8,001	7,051
+ 7,843	− 3,157	+ 2,368	− 5,553
10,955	**1,046**	**10,369**	**1,498**

4,609	8,631	5,662	7,368
− 4,609	− 4,554	− 4,554	− 5,601
0	**4,077**	**1,108**	**1,767**

70,700	52,581	86,264
− 69,702	+ 52,581	+ 21,299
998	**105,162**	**107,563**

88,846	74,116	96,869
+ 71,231	− 21,116	+ 76,867
160,077	**53,000**	**173,736**

83,679	75,499	59,698
− 22,969	+ 22,360	− 22,110
60,710	**97,859**	**37,588**

Time: _____ min.

Algebra Ready — Skill Practice 2 +/−

Name _____
Date _____
Period _____

Addition/Subtraction I

11,111 + 22,222 **33,333**	87,965 − 65,423 **22,542**	66,665 + 55,556 **122,221**
666,665 + 666,542 **1,333,207**	373,113 − 159,205 **213,908**	96,302 + 58,741 **155,043**
78,945 − 65,498 **13,447**	45,687 + 23,235 **68,922**	745,874 + 852,145 **1,598,019**
588,757 − 288,469 **300,288**	37,329 − 21,674 **15,655**	99,999 − 65,423 **34,576**
45,871 + 45,821 **91,692**	777,556 + 222,659 **1,000,215**	463,136 − 153,216 **309,920**
93,756 − 84,867 **8,889**	99,875 − 65,423 **34,452**	23,009 + 23,569 **46,578**
123,456 + 123,456 **246,912**	666,000 − 245,000 **421,000**	37,231 − 32,466 **4,765**

Time: _____ min.

Addition/Subtraction II

45,698 − 32,156 **13,542**	20,153 + 23,562 **43,715**	445,500 + 223,203 **668,703**
881,019 − 447,609 **433,410**	86,899 − 36,999 **49,900**	65,498 + 32,556 **98,054**
77,412 + 88,456 **165,868**	112,546 + 456,789 **569,335**	477,588 − 477,578 **10**
85,633 − 45,623 **40,010**	32,916 − 25,555 **7,361**	10,102 + 20,203 **30,305**
333,000 − 222,111 **110,889**	596,832 − 321,764 **275,068**	32,564 − 21,548 **11,016**
55,555 − 32,165 **23,390**	33,333 + 66,669 **100,002**	455,321 − 396,333 **58,988**
569,862 + 321,764 **891,626**	766,999 − 345,777 **421,222**	47,342 + 22,577 **69,919**

Time: _____ min.

Addition/Subtraction III

11,999	78,965	88,665
+ 22,777	− 56,423	+ 33,556
34,776	**22,542**	**122,221**

444,665	737,113	69,302
+ 555,542	− 591,205	+ 85,741
1,000,207	**145,908**	**155,043**

87,945	54,687	547,874
− 56,498	+ 32,235	+ 258,145
31,447	**86,922**	**806,019**

885,757	73,329	55,999
− 882,469	− 12,674	− 43,423
3,288	**60,655**	**12,576**

54,871	555,556	364,136
+ 54,821	+ 111,659	− 351,216
109,692	**667,215**	**12,920**

71,756	88,875	32,009
− 62,867	− 56,423	+ 32,569
8,889	**32,452**	**64,578**

321,456	555,000	73,231
+ 321,456	− 542,000	− 23,466
642,912	**13,000**	**49,765**

Time: _____ min.

Addition/Subtraction IV

54,698	30,153	544,500
− 23,156	+ 43,562	+ 322,203
31,542	**73,715**	**866,703**

771,019	68,899	56,498
− 337,609	− 63,999	+ 23,556
433,410	**4,900**	**80,054**

99,412	211,546	774,688
+ 66,456	+ 654,789	− 774,578
165,868	**866,335**	**110**

58,633	52,916	90,102
− 54,623	− 52,555	+ 21,203
4,010	**361**	**111,305**

777,000	695,832	23,564
− 444,111	− 123,764	− 20,548
332,889	**572,068**	**3,016**

88,888	36,000	554,321
− 23,165	+ 22,999	+ 693,333
65,723	**58,999**	**1,247,654**

965,862	667,999	74,342
+ 213,764	− 543,777	+ 11,577
1,179,626	**124,222**	**85,919**

Time: _____ min.

Algebra Ready — Skill Practice 1 ×

Name _____
Date _____
Period _____

Multiplication I

1 × 3 = **3**	2 × 2 = **4**	4 × 5 = **20**	3 × 6 = **18**
7 × 0 = **0**	5 × 9 = **45**	6 × 8 = **48**	9 × 3 = **27**
11 × 9 = **99**	4 × 4 = **16**	5 × 3 = **15**	9 × 7 = **63**
5 × 2 = **10**	3 × 7 = **21**	12 × 4 = **48**	10 × 2 = **20**
6 × 2 = **12**	7 × 4 = **28**	7 × 8 = **56**	3 × 5 = **15**
12 × 2 = **24**	5 × 6 = **30**	11 × 11 = **121**	9 × 0 = **0**
3 × 8 = **24**	12 × 6 = **72**	4 × 7 = **28**	8 × 4 = **32**

Time: _____ min.

Multiplication II

1 × 5 = **5**	8 × 8 = **64**	10 × 7 = **70**	6 × 6 = **36**
11 × 12 = **132**	0 × 3 = **0**	2 × 3 = **6**	12 × 5 = **60**
4 × 4 = **16**	7 × 11 = **77**	4 × 6 = **24**	7 × 2 = **14**
3 × 3 = **9**	10 × 10 = **100**	2 × 4 = **8**	12 × 8 = **96**
11 × 10 = **110**	5 × 5 = **25**	12 × 9 = **108**	9 × 3 = **27**
11 × 5 = **55**	6 × 1 = **6**	5 × 8 = **40**	7 × 7 = **49**
12 × 3 = **36**	20 × 9 = **180**	11 × 8 = **88**	10 × 7 = **70**

Time: _____ min.

Multiplication III

37 × 2 = **74**	51 × 30 = **1,530**	10 × 5 = **50**	62 × 9 = **558**
61 × 4 = **244**	17 × 14 = **238**	13 × 1 = **13**	13 × 12 = **156**
74 × 21 = **1,554**	37 × 33 = **1,221**	74 × 122 = **9,028**	48 × 23 = **1,104**
8 × 23 = **184**	90 × 2 = **180**	10 × 8 = **80**	51 × 34 = **1,734**
6 × 2 = **12**	62 × 18 = **1,116**	70 × 11 = **770**	15 × 4 = **60**
82 × 14 = **1,148**	14 × 14 = **196**	72 × 6 = **432**	55 × 23 = **1,265**
44 × 44 = **1,936**	10 × 4 = **40**	10 × 10 = **100**	55 × 33 = **1,815**

Time: _____ min.

Multiplication IV

47 × 24 = **1,128**	83 × 14 = **1,162**	87 × 11 = **957**	34 × 10 = **340**
11 × 12 = **132**	15 × 6 = **90**	86 × 15 = **1,290**	10 × 7 = **70**
78 × 23 = **1,794**	51 × 48 = **2,448**	63 × 17 = **1,071**	7 × 3 = **21**
38 × 11 = **418**	9 × 3 = **27**	11 × 12 = **132**	19 × 4 = **76**
46 × 2 = **92**	18 × 11 = **198**	46 × 18 = **828**	51 × 4 = **204**
66 × 14 = **924**	24 × 6 = **144**	27 × 9 = **243**	76 × 8 = **608**
35 × 44 = **1,540**	30 × 10 = **300**	112 × 99 = **11,088**	15 × 76 = **1,140**

Time: _____ min.

Algebra Ready — Skill Practice 2   ×

Name _____

Date _____

Period _____

Multiplication I

2	3	1	10
×5	×2	×3	×12
10	**6**	**3**	**120**
2	7	9	5
×9	×1	×5	×7
18	**7**	**45**	**35**
10	5	6	7
×3	×9	×9	×6
30	**45**	**54**	**42**
8	1	7	3
×9	×8	×12	×12
72	**8**	**84**	**36**
12	4	6	12
×8	×10	×8	×2
96	**40**	**48**	**24**
2	10	9	12
×7	×5	×7	×12
14	**50**	**63**	**144**
3	5	12	0
×1	×2	×3	×0
3	**10**	**36**	**0**

Time: _____ min.

Multiplication II

2	3	11	9
×8	×7	×0	×6
16	**21**	**0**	**54**
5	6	4	5
×8	×0	×6	×5
40	**0**	**24**	**25**
4	9	12	3
×4	×9	×10	×5
16	**81**	**120**	**15**
5	6	3	11
×9	×5	×6	×2
45	**30**	**18**	**22**
12	8	11	6
×6	×2	×1	×8
72	**16**	**11**	**48**
4	7	5	7
×1	×7	×10	×11
4	**49**	**50**	**77**
12	0	10	10
×3	×9	×8	×7
36	**0**	**80**	**70**

Time: _____ min.

Multiplication III

3 × 6 = **18**	4 × 1 = **4**	2 × 2 = **4**	11 × 13 = **143**
1 × 10 = **10**	6 × 2 = **12**	8 × 6 = **48**	4 × 8 = **32**
11 × 2 = **22**	6 × 10 = **60**	7 × 8 = **56**	6 × 7 = **42**
7 × 10 = **70**	2 × 9 = **18**	6 × 11 = **66**	4 × 10 = **40**
11 × 7 = **77**	5 × 9 = **45**	7 × 7 = **49**	13 × 1 = **13**
1 × 8 = **8**	9 × 6 = **54**	8 × 8 = **64**	13 × 13 = **169**
4 × 2 = **8**	6 × 3 = **18**	11 × 9 = **99**	1 × 9 = **9**

Time: _____ min.

Multiplication IV

4 × 6 = **24**	5 × 5 = **25**	13 × 2 = **26**	2 × 4 = **8**
7 × 6 = **42**	8 × 8 = **64**	6 × 0 = **0**	7 × 3 = **21**
6 × 2 = **12**	1 × 7 = **7**	10 × 12 = **120**	5 × 3 = **15**
7 × 0 = **0**	8 × 3 = **24**	5 × 4 = **20**	13 × 0 = **0**
14 × 0 = **0**	10 × 1 = **10**	13 × 8 = **104**	8 × 6 = **48**
6 × 8 = **48**	9 × 5 = **45**	7 × 8 = **56**	9 × 9 = **81**
14 × 1 = **14**	2 × 7 = **14**	12 × 6 = **72**	12 × 5 = **60**

Time: _____ min.

Algebra Ready — Skill Practice 3 ×

Name _____

Date _____

Period _____

Multiplication I

44 × 18 = **792**	47 × 20 = **940**	55 × 55 = **3,025**	70 × 35 = **2,450**
2,090 × 100 = **209,000**	90 × 100 = **9,000**	963 × 100 = **96,300**	45 × 100 = **4,500**
25 × 17 = **425**	75 × 43 = **3,225**	60 × 34 = **2,040**	32 × 25 = **800**
31 × 18 = **558**	6,665 × 100 = **666,500**	1,000 × 687 = **687,000**	1,000 × 45 = **45,000**
153 × 100 = **15,300**	3,009 × 1,000 = **3,009,000**	7,412 × 100 = **741,200**	10 × 20 = **200**

Time: _____ min.

Multiplication II

12 × 56 = **672**	44 × 22 = **968**	11 × 45 = **495**	60 × 45 = **2,700**
3,090 × 100 = **309,000**	100 × 110 = **11,000**	863 × 90 = **77,670**	55 × 100 = **5,500**
35 × 27 = **945**	85 × 33 = **2,805**	70 × 44 = **3,080**	42 × 35 = **1,470**
51 × 28 = **1,428**	7,665 × 100 = **766,500**	1,000 × 587 = **587,000**	1,000 × 325 = **325,000**
253 × 100 = **25,300**	6,009 × 1,000 = **6,009,000**	2,412 × 100 = **241,200**	900 × 20 = **18,000**

Time: _____ min.

Multiplication III

39	38	65	29
×76	×16	×33	×92
2,964	**608**	**2,145**	**2,668**

2,909	40	369	54
×99	×100	×100	×100
287,991	**4,000**	**36,900**	**5,400**

777	6,500	99	13
×100	×10	×18	×25
77,700	**65,000**	**1,782**	**325**

52	57	60	23
×17	×43	×43	×25
884	**2,451**	**2,580**	**575**

351	9,003	4,721	88
×100	×1,000	×100	×20
35,100	**9,003,000**	**472,100**	**1,760**

Time: _____ min.

Multiplication IV

99	98	47	65
×66	×24	×45	×22
6,534	**2,352**	**2,115**	**1,430**

9,303	1,100	368	63
×100	×110	×90	×100
930,300	**121,000**	**33,120**	**6,300**

855	6,400	109	14
×90	×20	×19	×51
76,950	**128,000**	**2,071**	**714**

53	58	77	24
×27	×33	×44	×35
1,431	**1,914**	**3,388**	**840**

352	9,660	4,212	999
×100	×1,000	×100	×30
35,200	**9,660,000**	**421,200**	**29,970**

Time: _____ min.

Algebra Ready — Skill Practice 4   ×

Name _____

Date _____

Period _____

Multiplication I

81 × 10 = **810**	50 × 17 = **850**	35 × 35 = **1,225**	631 × 10 = **6,310**
52 × 10 = **520**	100 × 10 = **1,000**	625 × 100 = **62,500**	875 × 10 = **8,750**
76 × 50 = **3,800**	25 × 12 = **300**	89 × 70 = **6,230**	90 × 31 = **2,790**
7,599 × 10 = **75,990**	43 × 10 = **430**	6,594 × 10 = **65,940**	549 × 10 = **5,490**
400 × 100 = **40,000**	90 × 55 = **4,950**	22 × 17 = **374**	51 × 49 = **2,499**

Time: _____ min.

Multiplication II

5,003 × 100 = **500,300**	601 × 100 = **60,100**	5,180 × 100 = **518,000**	4,409 × 100 = **440,900**
24 × 100 = **2,400**	3,227 × 100 = **322,700**	8,070 × 100 = **807,000**	100 × 45 = **4,500**
3,090 × 70 = **216,300**	505 × 100 = **50,500**	860 × 200 = **172,000**	255 × 100 = **25,500**
100 × 20 = **2,000**	80 × 33 = **2,640**	100 × 44 = **4,400**	40 × 30 = **1,200**
57 × 69 = **3,933**	7,000 × 77 = **539,000**	1,000 × 100 = **100,000**	1,000 × 303 = **303,000**

Time: _____ min.

Multiplication III

63 × 10 = **630**	50 × 35 = **1,750**	53 × 35 = **1,855**	453 × 10 = **4,530**
34 × 10 = **340**	211 × 10 = **2,110**	443 × 100 = **44,300**	693 × 10 = **6,930**
58 × 50 = **2,900**	43 × 12 = **516**	66 × 70 = **4,620**	72 × 31 = **2,232**
8,333 × 10 = **83,330**	61 × 10 = **610**	5,312 × 10 = **53,120**	721 × 10 = **7,210**
91 × 25 = **2,275**	80 × 23 = **1,840**	58 × 25 = **1,450**	71 × 22 = **1,562**

Time: _____ min.

Multiplication IV

9,300 × 100 = **930,000**	883 × 100 = **88,300**	3,362 × 100 = **336,200**	2,681 × 100 = **268,100**
86 × 100 = **8,600**	1,409 × 100 = **140,900**	6,252 × 100 = **625,200**	100 × 27 = **2,700**
1,272 × 70 = **89,040**	323 × 100 = **32,300**	1,042 × 200 = **208,400**	873 × 100 = **87,300**
100 × 33 = **3,300**	62 × 33 = **2,046**	100 × 68 = **6,800**	50 × 30 = **1,500**
404 × 100 = **40,400**	4,848 × 1,000 = **4,848,000**	8,682 × 100 = **868,200**	100 × 22 = **2,200**

Time: _____ min.

Algebra Ready — Skill Practice 5 ×

Name _____
Date _____
Period _____

Multiplication I

93 ×76 **7,068**	38 ×61 **2,318**	56 ×33 **1,848**	29 ×29 **841**
700 ×100 **70,000**	5,600 × 10 **56,000**	99 ×81 **8,019**	31 ×25 **775**
21 ×46 **966**	13 ×10 **130**	9,000 × 10 **90,000**	36 ×10 **360**
73 ×25 **1,825**	62 ×23 **1,426**	30 ×25 **750**	55 ×22 **1,210**
29 ×31 **899**	76 ×45 **3,420**	17 ×33 **561**	84 ×75 **6,300**

Time: _____ min.

Multiplication II

33 ×66 **2,178**	98 ×42 **4,116**	74 ×45 **3,330**	56 ×22 **1,232**
800 ×90 **72,000**	4,600 × 20 **92,000**	101 ×91 **9,191**	41 ×15 **615**
200 ×20 **4,000**	1,666 × 10 **16,660**	900 ×10 **9,000**	42 ×12 **504**
222 ×100 **22,200**	6,666 ×1,000 **6,666,000**	2,400 ×100 **240,000**	100 ×39 **3,900**
41 ×64 **2,624**	57 ×21 **1,197**	24 ×32 **768**	67 ×13 **871**

Time: _____ min.

Multiplication III

44	74	88	70
×81	×20	×55	×53
3,564	**1,480**	**4,840**	**3,710**

13	5,666	1,000	1,000
×18	×100	×786	×444
234	**566,600**	**786,000**	**444,000**

49	31	1,999	54
×46	×10	×10	×10
2,254	**310**	**19,990**	**540**

48	99	40	73
×100	×55	×71	×49
4,800	**5,445**	**2,840**	**3,577**

35	64	27	19
×53	×46	×72	×91
1,855	**2,944**	**1,944**	**1,729**

Time: _____ min.

Multiplication IV

21	88	11	2,681
×56	×22	×22	×100
1,176	**1,936**	**242**	**268,100**

15	5,776	1,000	1,000
×28	×100	×785	×523
420	**577,600**	**785,000**	**523,000**

600	9,848	182	24
×20	×10	×10	×21
12,000	**98,480**	**1,820**	**504**

39	3,888	4,500	1,000
×69	×10	×100	×888
2,691	**38,880**	**450,000**	**888,000**

37	84	71	41
×14	×25	×31	×76
518	**2,100**	**2,201**	**3,116**

Time: _____ min.

Algebra Ready — Skill Practice 6 ×

Name _____

Date _____

Period _____

Multiplication I

342	626	201
× 600	× 323	× 642
205,200	**202,198**	**129,042**

263	115	649
× 298	× 115	× 102
78,374	**13,225**	**66,198**

625	111	701
× 525	× 842	× 701
328,125	**93,462**	**491,401**

845	868	498
× 230	× 868	× 369
194,350	**753,424**	**183,762**

111	302	329
× 222	× 741	× 674
24,642	**223,782**	**221,746**

Time: _____ min.

Multiplication II

903	601	188
× 700	× 800	× 188
632,100	**480,800**	**35,344**

224	227	770
× 111	× 800	× 199
24,864	**181,600**	**153,230**

390	505	160
× 670	× 245	× 290
261,300	**123,725**	**46,400**

340	666	920
× 980	× 710	× 410
333,200	**472,860**	**377,200**

721	890	760
× 465	× 313	× 244
335,265	**278,570**	**185,440**

Time: _____ min.

Multiplication III

442 × 700 **309,400**	726 × 423 **307,098**	301 × 742 **223,342**
363 × 198 **71,874**	315 × 915 **288,225**	849 × 302 **256,398**
825 × 725 **598,125**	311 × 142 **44,162**	901 × 501 **451,401**
645 × 430 **277,350**	168 × 668 **112,224**	698 × 569 **397,162**
211 × 322 **67,942**	402 × 841 **338,082**	429 × 774 **332,046**

Time: _____ min.

Multiplication IV

703 × 900 **632,700**	801 × 600 **480,600**	988 × 288 **284,544**
524 × 711 **372,564**	427 × 500 **213,500**	170 × 799 **135,830**
990 × 170 **168,300**	605 × 145 **87,725**	260 × 190 **49,400**
440 × 880 **387,200**	766 × 610 **467,260**	420 × 910 **382,200**
921 × 165 **151,965**	390 × 813 **317,070**	460 × 744 **342,240**

Time: _____ min.

Algebra Ready — Skill Practice 1 ÷

Name _____
Date _____
Period _____

Division I

$1\overline{)6} = 6$ $\quad 2\overline{)12} = 6$ $\quad 6\overline{)36} = 6$ $\quad 9\overline{)27} = 3$

$5\overline{)20} = 4$ $\quad 5\overline{)35} = 7$ $\quad 5\overline{)15} = 3$ $\quad 5\overline{)10} = 2$

$4\overline{)28} = 7$ $\quad 3\overline{)27} = 9$ $\quad 4\overline{)32} = 8$ $\quad 3\overline{)18} = 6$

$8\overline{)56} = 7$ $\quad 7\overline{)42} = 6$ $\quad 6\overline{)24} = 4$ $\quad 9\overline{)54} = 6$

$10\overline{)40} = 4$ $\quad 8\overline{)80} = 10$ $\quad 10\overline{)70} = 7$ $\quad 10\overline{)10} = 1$

$7\overline{)21} = 3$ $\quad 4\overline{)16} = 4$ $\quad 8\overline{)64} = 8$ $\quad 7\overline{)49} = 7$

$6\overline{)66} = 11$ $\quad 3\overline{)33} = 11$ $\quad 2\overline{)20} = 10$ $\quad 6\overline{)60} = 10$

Time: _____ min.

Division II

$8\overline{)8} = 1$ $\quad 1\overline{)7} = 7$ $\quad 7\overline{)7} = 1$ $\quad 4\overline{)12} = 3$

$9\overline{)81} = 9$ $\quad 9\overline{)36} = 4$ $\quad 9\overline{)18} = 2$ $\quad 9\overline{)72} = 8$

$5\overline{)40} = 8$ $\quad 5\overline{)5} = 1$ $\quad 5\overline{)30} = 6$ $\quad 5\overline{)25} = 5$

$3\overline{)9} = 3$ $\quad 3\overline{)24} = 8$ $\quad 2\overline{)8} = 4$ $\quad 8\overline{)16} = 2$

$10\overline{)20} = 2$ $\quad 10\overline{)50} = 5$ $\quad 8\overline{)80} = 10$ $\quad 7\overline{)70} = 10$

$6\overline{)12} = 2$ $\quad 7\overline{)28} = 4$ $\quad 2\overline{)18} = 9$ $\quad 8\overline{)48} = 6$

$9\overline{)99} = 11$ $\quad 5\overline{)55} = 11$ $\quad 7\overline{)77} = 11$ $\quad 10\overline{)30} = 3$

Time: _____ min.

Division III

$7\overline{)0} = 0 \quad 1\overline{)11} = 11 \quad 2\overline{)0} = 0 \quad 1\overline{)8} = 8$

$8\overline{)24} = 3 \quad 3\overline{)15} = 5 \quad 4\overline{)28} = 7 \quad 8\overline{)32} = 4$

$7\overline{)56} = 8 \quad 6\overline{)54} = 9 \quad 2\overline{)16} = 8 \quad 4\overline{)36} = 9$

$2\overline{)4} = 2 \quad 2\overline{)14} = 7 \quad 6\overline{)48} = 8 \quad 2\overline{)10} = 5$

$8\overline{)40} = 5 \quad 6\overline{)30} = 5 \quad 4\overline{)20} = 5 \quad 9\overline{)45} = 5$

$6\overline{)66} = 11 \quad 2\overline{)22} = 11 \quad 4\overline{)44} = 11 \quad 8\overline{)88} = 11$

$10\overline{)10} = 1 \quad 10\overline{)30} = 3 \quad 10\overline{)70} = 7 \quad 6\overline{)60} = 10$

Time: _____ min.

Division IV

$9\overline{)90} = 10 \quad 77\overline{)0} = 0 \quad 11\overline{)66} = 6 \quad 10\overline{)90} = 9$

$6\overline{)42} = 7 \quad 4\overline{)24} = 6 \quad 7\overline{)14} = 2 \quad 9\overline{)63} = 7$

$2\overline{)6} = 3 \quad 3\overline{)12} = 4 \quad 4\overline{)8} = 2 \quad 2\overline{)10} = 5$

$6\overline{)18} = 3 \quad 5\overline{)45} = 9 \quad 2\overline{)6} = 3 \quad 3\overline{)21} = 7$

$9\overline{)45} = 5 \quad 9\overline{)63} = 7 \quad 8\overline{)56} = 7 \quad 9\overline{)54} = 6$

$8\overline{)72} = 9 \quad 4\overline{)72} = 18 \quad 8\overline{)56} = 7 \quad 4\overline{)56} = 14$

$2\overline{)48} = 24 \quad 6\overline{)72} = 12 \quad 3\overline{)63} = 21 \quad 4\overline{)56} = 14$

Time: _____ min.

Algebra Ready — Skill Practice 2 ÷

Name _____
Date _____
Period _____

Division I

$9\overline{)63} = 7$ $9\overline{)81} = 9$ $9\overline{)72} = 8$

$6\overline{)42} = 7$ $6\overline{)43} = 7R1$ $6\overline{)45} = 7R3$

$6\overline{)24} = 4$ $6\overline{)26} = 4R2$ $6\overline{)27} = 4R3$

$2\overline{)36} = 18$ $2\overline{)37} = 18R1$ $2\overline{)38} = 19$

$8\overline{)49} = 6R1$ $7\overline{)56} = 8$ $5\overline{)30} = 6$

$6\overline{)54} = 9$ $5\overline{)28} = 5R3$ $4\overline{)67} = 16R3$

Time: _____ min.

Division II

$4\overline{)36} = 9$ $7\overline{)49} = 7$ $3\overline{)24} = 8$

$3\overline{)66} = 22$ $2\overline{)34} = 17$ $4\overline{)84} = 21$

$10\overline{)50} = 5$ $10\overline{)52} = 5R2$ $5\overline{)53} = 10R3$

$7\overline{)63} = 9$ $7\overline{)65} = 9R2$ $9\overline{)65} = 7R2$

$2\overline{)68} = 34$ $3\overline{)72} = 24$ $2\overline{)242} = 121$

$2\overline{)17} = 8R1$ $8\overline{)52} = 6R4$ $7\overline{)30} = 4R2$

Time: _____ min.

Division III

$$\begin{array}{r}8\\8\overline{)64}\end{array} \quad \begin{array}{r}4\\7\overline{)28}\end{array} \quad \begin{array}{r}7\\6\overline{)42}\end{array}$$

$$\begin{array}{r}4\\9\overline{)36}\end{array} \quad \begin{array}{r}4R2\\9\overline{)38}\end{array} \quad \begin{array}{r}4R5\\9\overline{)41}\end{array}$$

$$\begin{array}{r}5\\9\overline{)45}\end{array} \quad \begin{array}{r}5R3\\9\overline{)48}\end{array} \quad \begin{array}{r}5R7\\9\overline{)52}\end{array}$$

$$\begin{array}{r}7\\5\overline{)35}\end{array} \quad \begin{array}{r}8R1\\8\overline{)65}\end{array} \quad \begin{array}{r}9\\7\overline{)63}\end{array}$$

$$\begin{array}{r}13\\4\overline{)52}\end{array} \quad \begin{array}{r}43R1\\2\overline{)87}\end{array} \quad \begin{array}{r}15\\5\overline{)75}\end{array}$$

$$\begin{array}{r}111\\2\overline{)222}\end{array} \quad \begin{array}{r}111R1\\2\overline{)223}\end{array} \quad \begin{array}{r}102\\6\overline{)612}\end{array}$$

Time: _____ min.

Division IV

$$\begin{array}{r}4\\8\overline{)32}\end{array} \quad \begin{array}{r}7\\8\overline{)56}\end{array} \quad \begin{array}{r}6\\3\overline{)18}\end{array}$$

$$\begin{array}{r}17\\4\overline{)68}\end{array} \quad \begin{array}{r}81\\3\overline{)243}\end{array} \quad \begin{array}{r}14\\10\overline{)140}\end{array}$$

$$\begin{array}{r}6R2\\8\overline{)50}\end{array} \quad \begin{array}{r}8R4\\8\overline{)68}\end{array} \quad \begin{array}{r}9R2\\10\overline{)92}\end{array}$$

$$\begin{array}{r}234\\2\overline{)468}\end{array} \quad \begin{array}{r}131\\4\overline{)524}\end{array} \quad \begin{array}{r}6\\6\overline{)36}\end{array}$$

$$\begin{array}{r}5R3\\7\overline{)38}\end{array} \quad \begin{array}{r}6R3\\9\overline{)57}\end{array} \quad \begin{array}{r}9R3\\8\overline{)75}\end{array}$$

$$\begin{array}{r}7\\11\overline{)77}\end{array} \quad \begin{array}{r}7R2\\11\overline{)79}\end{array} \quad \begin{array}{r}11R2\\8\overline{)90}\end{array}$$

Time: _____ min.

Algebra Ready — Skill Practice 3 ÷

Name _____
Date _____
Period _____

Division I

$8\overline{)48}=6 \qquad 2\overline{)14}=7 \qquad 9\overline{)63}=7$

$2\overline{)6}=3 \qquad 5\overline{)40}=8 \qquad 6\overline{)54}=9$

$12\overline{)72}=6 \qquad 14\overline{)84}=6 \qquad 17\overline{)68}=4$

$12\overline{)74}=6R2 \qquad 14\overline{)89}=6R5 \qquad 17\overline{)78}=4R10$

$4\overline{)36}=9 \qquad 4\overline{)39}=9R3 \qquad 5\overline{)37}=7R2$

$10\overline{)240}=24 \qquad 24\overline{)240}=10 \qquad 12\overline{)243}=20R3$

Time: _____ min.

Division II

$4\overline{)16}=4 \qquad 7\overline{)56}=8 \qquad 7\overline{)28}=4$

$9\overline{)18}=2 \qquad 9\overline{)21}=2R3 \qquad 8\overline{)72}=9$

$19\overline{)76}=4 \qquad 3\overline{)51}=17 \qquad 2\overline{)84}=42$

$10\overline{)87}=8R7 \qquad 7\overline{)63}=9 \qquad 7\overline{)66}=9R3$

$9\overline{)75}=8R3 \qquad 7\overline{)24}=3R3 \qquad 8\overline{)58}=7R2$

$4\overline{)844}=211 \qquad 3\overline{)243}=81 \qquad 4\overline{)847}=211R3$

Time: _____ min.

Division III

$$5\overline{)15}=3 \qquad 4\overline{)32}=8 \qquad 3\overline{)27}=9$$

$$4\overline{)12}=3 \qquad 4\overline{)14}=3R2 \qquad 8\overline{)18}=2R2$$

$$2\overline{)64}=32 \qquad 32\overline{)64}=2 \qquad 32\overline{)74}=2R10$$

$$7\overline{)329}=47 \qquad 8\overline{)968}=121 \qquad 6\overline{)516}=86$$

$$11\overline{)88}=8 \qquad 11\overline{)92}=8R4 \qquad 11\overline{)79}=7R2$$

$$10\overline{)70}=7 \qquad 5\overline{)70}=14 \qquad 5\overline{)74}=14R4$$

Time: _____ min.

Division IV

$$3\overline{)24}=8 \qquad 6\overline{)48}=8 \qquad 5\overline{)25}=5$$

$$9\overline{)81}=9 \qquad 9\overline{)83}=9R2 \qquad 3\overline{)10}=3R1$$

$$8\overline{)424}=53 \qquad 4\overline{)980}=245 \qquad 7\overline{)364}=52$$

$$8\overline{)426}=53R2 \qquad 4\overline{)982}=245R2 \qquad 12\overline{)96}=8$$

$$11\overline{)99}=9 \qquad 11\overline{)104}=9R5 \qquad 11\overline{)110}=10$$

$$3\overline{)12}=4 \qquad 22\overline{)88}=4 \qquad 4\overline{)627}=156R3$$

Time: _____ min.

Algebra Ready — Skill Practice 4 ÷

Name _____
Date _____
Period _____

Division I

19
$2\overline{)38}$

$2R4$
$19\overline{)42}$

25
$3\overline{)75}$

$14R2$
$4\overline{)58}$

$12R5$
$7\overline{)89}$

17
$3\overline{)51}$

137
$6\overline{)822}$

$137R5$
$6\overline{)827}$

133
$5\overline{)665}$

12
$10\overline{)120}$

55
$10\overline{)550}$

111
$7\overline{)777}$

$1{,}231$
$3\overline{)3{,}693}$

144
$6\overline{)864}$

$152R2$
$3\overline{)458}$

Time: _____ min.

Division II

$8R3$
$7\overline{)59}$

$21R3$
$4\overline{)87}$

135
$4\overline{)540}$

500
$5\overline{)2{,}500}$

312
$3\overline{)936}$

19
$4\overline{)76}$

13
$12\overline{)156}$

22
$16\overline{)352}$

$139R6$
$7\overline{)979}$

$611R3$
$6\overline{)3{,}669}$

148
$4\overline{)592}$

16
$2\overline{)32}$

$42R1$
$2\overline{)85}$

49
$7\overline{)343}$

121
$8\overline{)968}$

Time: _____ min.

Division III

9R2 3)29	8R3 6)51	7R2 7)51
47 7)329	22 6)132	12R4 10)124
90R4 5)454	44 4)176	323 3)969
772 7)5,404	167R3 4)671	16R3 4)67
20R4 15)304	196R3 5)983	47 20)940

Time: _____ min.

Division IV

9R3 8)75	5R2 3)17	6R2 11)68
18R1 3)55	67R2 7)471	81R4 6)490
10 33)330	14 12)168	289R1 2)579
100 33)3,300	54R1 3)163	239 2)478
1,000 33)33,000	76 13)988	112R3 4)451

Time: _____ min.

Algebra Ready   Additional Word Problem Warm-Ups   WP

Additional Word Problem Warm-Ups

1. One truck is carrying five tanks weighing 2 tons each. A second truck is carrying three tanks weighing 5 tons each. Altogether, how much do the tanks weigh?

   25 tons

2. An investor bought 14 acres of land for $80,000. She later subdivided the land into 22 lots that she sold for $4,750 apiece. What was her profit on the sale?

   $24,500

3. Jose's patio is 5 square yards. If he covers 1 square yard with bricks, what fraction of the patio will not have bricks?

   $\frac{4}{5}$

4. Jake's father is 55 years old. He is 15 years more than twice Jake's age. How old is Jake? Write an equation for this problem, then solve it.

   Jake is 20.

   $2x + 15 = 55$

5. A girl has to be at school by 7:55 A.M. and it takes her 15 minutes to get dressed, 30 minutes to eat, and 35 minutes to walk to school. What time should she get up?

6:35 A.M.

6. Sandra drives her car 52 miles every workday to get to and from work. She works five days a week. How many miles does she commute each week?

260 miles

7. Replace each blank with the correct digit.

$$\begin{array}{r} 4,3\_2 \\ 45\_ \\ +\_,127 \\ \hline 8,893 \end{array} \qquad \begin{array}{r} 4,3①2 \\ 45④ \\ +④,127 \\ \hline 8,893 \end{array}$$

8. A total of 28 handshakes were exchanged at a party. Assuming that each person shakes hands with all the rest once only, how many people attended the party?

8 people

Algebra Ready   Additional Word Problem Warm-Ups   WP

9. If Tasharra puts $3.00 in the bank on January 1, $6.00 on February 1, $9.00 on March 1, $12.00 on April 1, and so on, how much money would she save in one year?

$234.00

10. If Marko's brother Kevin puts $3.00 in the bank on January 1, $6.00 on February 1, $12.00 on March 1, $24.00 on April 1, and so on, how much money would he save in one year?

$12,285.00

11. A package of Chris's favorite cheese is on sale for $1.97. In order to make her favorite recipe, she needs nine packages. How much will Chris pay?

$17.73

12. Palindromic numbers read the same backward and forward. For example, 1221 and 696 are palindromic numbers. What is the next year that is a palindromic number?

2112

13. In its first year, a school musical group sold 1,500 tickets. In its second year, it sold 1,750 tickets. In its third year, it sold 152 less than in its second year. How many tickets were sold altogether in three years?

4,848 tickets

14. Ethel made a list of all the whole numbers between 1 and 100. How many times did she write the digit 2?

20 times

15. How many seconds are in an hour?

3,600 seconds

16. Michael wanted an allowance. His father gave him a choice of being paid on a weekly or on a daily basis. He said he would either pay him $2.25 a week or pay him in the following manner each week: on Monday he would give him $0.02; on Tuesday $0.04; on Wednesday $0.08; and on through Sunday. Which choice would give Michael more money?

Second choice

Algebra Ready — Additional Word Problem Warm-Ups   WP

17. Bobby has five times as many apples as Marty, who has 3. How many apples does Bobby have?

Bobby has 15 apples.

18. Maria's garden is 20 feet long and 10 feet wide. How many square feet are in her garden? If she wanted to enlarge her garden by 50 square feet, how many feet longer should she make it?

200 square feet

5 feet

19. Four people share a car for a period of one year, and the mean number of miles traveled by each person is 163 per month. How many total miles will the four people travel in one year?

7,824 miles

20. In a survey of 300 ninth-grade students, 63% like rock music. How many students do not like rock music?

111 students do not like rock music.

21. Marge has three times as many baseball cards as Vicki, who has two times as many as Craig. If Craig has four baseball cards, how many does Marge have?

Marge has 24 baseball cards.

22. If Brian has four times as many posters as Suzy, and Suzy has one-fourth as many as Doug, who has eight, how many does Terry have if Terry has two more than Brian?

Terry has 10 posters.

23. Mountain Taxi Service charges $0.25 for the first mile and $0.10 for each additional mile. If Keith's cab fare is $2.35, how far did he ride?

22 miles

24. Naomi got 22 comic books from her friend who was moving and got twice as many from her friend's sister. She gave half of her comic books to her brother. Then she had to throw away 7 of the ones she had left because the dog chewed them up. How many comic books does she have now?

26 comic books

Algebra Ready　　　　　　　　　　Glossary　　　G

acute angle	An acute angle is an angle that measures less than 90°. That is, it is smaller than a right angle. The angle looks smaller than the letter L. ∠ABC and ∠DEF are examples of acute angles.

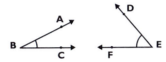

addend	An addend is any number or term being added to another. In the math sentence 3 + 6 = 9, the 3 and the 6 are both addends. In 2x + 3a, 2x and 3a are the addends.
adjacent angles	Adjacent angles are angles that are right next to each other and share a common side or ray. In the diagram, ∠1 and ∠2, ∠2 and ∠3, ∠3 and ∠4, and ∠4 and ∠1 are all pairs of adjacent angles.

algebra	Algebra is a part of mathematics that deals with variables, symbols, and numbers.
angle	An angle is a shape formed by two rays with a common vertex or point of origin.

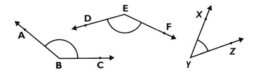

approximate (verb)	To approximate is to obtain an answer that is not exact but is close enough to serve the purpose. Approximating uses the skill of rounding. For example, five notebooks that have a price of $1.99 each would cost approximately $10.00 ($2.00 × 5).
area	The area is the space inside a closed two-dimensional figure. Area is labeled with square units.

associative property of addition	The associative property of addition allows you to group and add numbers in any order while getting the same sum. 3 + (5 + 8) = 16 and (3 + 5) + 8 = 16. In general, (a + b) + c = a + (b + c).
associative property of multiplication	The associative property of multiplication allows you to group and multiply numbers in any order while getting the same product. (3 · 4) · 2 = 24 and 3 · (4 · 2) = 24. In general, (a · b) · c = a · (b · c).
average	An average is the sum of the values of a set of items divided by the number of items added. The average of 2, 5, and 11 is found by adding 2 + 5 + 11, which is 18, and then dividing 18 by 3 because there are three numbers. The average is 6.
base	The base, in terms of exponents, is the number in front of the exponent. The base is multiplied by itself as many times as the exponent says. In 5^2, 5 is the base.
base	The base, in terms of geometry, generally refers to the side on which a figure rests.
binomial	A binomial is a polynomial with exactly two unlike terms. An example of a binomial is 3x + 2y. The x and y make the two terms unlike. Another example is 4 + 2d. A whole number is unlike a term with a variable. *Bi* is a prefix that means 2.
Cartesian graph	A Cartesian graph is a set of perpendicular number lines called the x-axis and y-axis that make a coordinate system. It can be used for locating points on a plane or for graphing equations. It is also called a coordinate graph or a coordinate system.

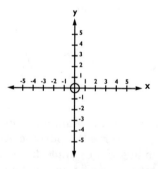

centi-	*Centi* is a metric prefix that means 1/100th (or 0.01). Think about pennies (cents) in $1.00 or centimeters in 1 meter. There are 100 in both.

centimeter	A centimeter is a small unit of metric length that is 1/100th of a meter. It is between $\frac{1}{4}$ and $\frac{1}{2}$ of an inch long.

⊢――⊣
1 cm

chord	A chord is a line segment that connects any two points of a circle. $\overline{AB}$ and $\overline{CD}$ are examples of chords.

circle	A circle is a closed shape without corners that has all points equal distance from the center.

circumference	The circumference is the boundary line of a circle. The formula for finding circumference is $C = 2\pi r$, or $C = \pi d$.
coefficient	The coefficient is the number directly in front of a variable. It is the number by which the variable is multiplied. In the binomial $3x + \frac{5}{8}y$, the 3 and the $\frac{5}{8}$ are the coefficients.
commutative property of addition	The commutative property of addition allows you to switch the positions of the addends (the numbers you add) while still getting the same sum. For example, $3 + 8 = 11$ and $8 + 3 = 11$. In general, the rule is $a + b = b + a$.
commutative property of multiplication	The commutative property of multiplication allows you to switch the positions of the numbers you multiply while still getting the same product. For example, $3 \cdot 8 = 24$ and $8 \cdot 3 = 24$. In general, the rule is $ab = ba$.
complementary angles	Complementary angles are angles that when added equal 90°. That is, complementary angles form a right angle.

composite number	A composite number is a positive whole number with more than two factors. For example, 4 is a composite number because 4 has more than two factors; 1, 2, and 4 all divide evenly into 4.
congruent	Any figures that have exactly the same size and shape are congruent. The symbol for showing that two angles, segments, or shapes are congruent is ≅.

consecutive numbers	Consecutive numbers are numbers that follow each other in a sequential pattern. For example, 1, 2, and 3 are consecutive whole numbers, and 2, 4, and 6 are consecutive even numbers.
coordinate	A coordinate is the "address" of a point or set of points on a line or graph. In this diagram, the coordinate of G is ⁻3.

counting numbers	Counting numbers are positive whole numbers: 1, 2, 3, 4, etc.
cubed	Cubed, in terms of exponents, refers to an exponent (power) of 3. 2^3 is stated two cubed, meaning 2 · 2 · 2, or 8.
data	Data are numerical information that can be put into charts, tables, etc., for comparison. Data are also the numerical information in a story problem that helps you solve it.
decimal	A decimal is any number written using a decimal point. It shows a whole number and a fractional part. **3.2** ← decimal ↑ **decimal point**
denominator	The denominator is the number below the line in a fraction. The denominator tells you how many equal pieces one whole has been divided into. In the fraction $\frac{3}{4}$, the 4 is the denominator, and there are 3 of 4 equal parts of one whole.

diameter	The diameter is a line segment across a circle that goes through the center and divides the circle into two equal halves. $\overline{AB}$ is the diameter of the circle below.

difference	The difference is the answer to a subtraction problem. For example, in the problem $8 - 5 = 3$, the difference is 3.
digit	A digit is any single whole number 0 through 9. For example, in the number 2,345, the digits are 2, 3, 4, and 5.
distribute	To distribute means to multiply each term within a set of grouping symbols by the term immediately preceding or following the grouping symbols. For example, $12(9 - 5)$ $12 \cdot 9 - 12 \cdot 5$ $108 - 60 = 48$
distributive property of multiplication over addition	The distributive property of multiplication over addition allows you to simplify a problem by sending a multiplier through a set of addends that are within grouping symbols. In general, $a(b + c) = ab + ac$. For example, $12(5 + 9)$ $12 \cdot 5 + 12 \cdot 9$ $60 + 108 = 168$
distributive property of multiplication over subtraction	The distributive property of multiplication over subtraction allows you to simplify a problem by sending a multiplier through a set of numbers within the grouping symbols that are being subtracted. In general, $a(b - c) = ab - ac$. Example: $12(9 - 5)$ $12 \cdot 9 - 12 \cdot 5$ $108 - 60 = 48$
divide	To divide means to split a number or number of items into parts of equal number. For example, Suzie wants to share six cookies with her two friends. There are a total of three people, and 6 divided by 3 is 2. She and her friends will each get two cookies.
dividend	In a division problem, the dividend is the number that is being split or divided into pieces. In the examples $\frac{12}{3}$, $12 \div 3$, and $3\overline{)12}$, the 12 is the dividend.

divisible	Divisible describes a number that can be divided by another number evenly with no remainder. For example, 6 is divisible by 2 ($6 \div 2 = 3$), but 7 is not divisible by 2 ($7 \div 2 = 3$ R1).
divisor	The divisor is the number that divides or splits another number into pieces. In the examples $\frac{12}{3}$, $12 \div 3$, and $3\overline{)12}$, the 3 is the divisor. The 12 is split into 3 equal pieces.
double	To double a number or term means to multiply the number by 2. For example, double (or twice) the number 2 is $2 \cdot 2$, which equals 4. To double 3y means $3y \cdot 2$, which equals 6y.
equation	An equation is a mathematical sentence that has two equivalent, or equal, expressions separated by an equal sign (=). There are numerical equations (those that have only numbers) and algebraic equations (those that have at least one variable).
equilateral triangle	An equilateral triangle is a triangle in which all three sides are equal in length. Each of the angles equals 60°. $\triangle ABC$ is an equilateral triangle.

equivalent	Equivalent refers to terms or expressions that have equal values. Equivalent terms may appear in different formats. $\frac{3}{4}$, $\frac{6}{8}$, 0.75, and 75% are all equivalent terms because they all have the same value.
estimate (verb)	To estimate is to make an educated guess based on information in a problem or to give an answer close to the exact number. Estimating often requires rounding. You could estimate that $28 \cdot 7$ is about 210 because 28 is close to 30 and $30 \cdot 7$ is 210.
evaluate	To evaluate is to find the value of an algebraic expression once the values of the variables are known. Example: Evaluate $a + b$ if $a = 4$ and $b = 2$. $a + b$ $4 + 2 = 6$

exponent	The exponent is the small raised number to the upper right of the base number that tells you how many times to multiply the base by itself. For example, in the term 2^3, 3 is the exponent and means you are to multiply 2 by itself three times. $2 \cdot 2 \cdot 2 = 8$, so $2^3 = 8$.
expression	An expression is a mathematical statement that stands for a given value. $2 + 3$ is a numerical expression. $a + b + 4$ is an algebraic expression.
factor (noun)	A factor is a number that divides evenly into another number. For example, the factors of 10 are 1, 2, 5, and 10.
factor (verb)	To factor a number or term is to find the numbers or terms that divide evenly into it.
formula	A formula is a recipe or equation used to find specific information. For example, the formula for the area of a rectangle is always Area = length × width (A = lw).
fraction	A fraction tells you what part of one whole unit you have. For example, $\frac{1}{2}$ means that you have one of two equal pieces that make one whole unit.
geometry	Geometry is the study of points, lines, angles, planes, and shapes, and their relationships.
graph (noun)	A graph is a visual display of data or other information. There are several different forms a graph can take: bar graph, line graph, and circle graph are some examples.
graph (verb)	To graph is to organize information in a visual format. You may use plotted points (coordinates), bars, or pieces of a circle to show the information.
greatest common factor (GCF)	The greatest common factor (GCF) is the largest factor common to two or more numbers. For example, the GCF of 12 (1, 2, 3, 4, 6, 12) and 20 (1, 2, 4, 5, 10, 20) is 4.
grouping symbols	Grouping symbols are symbols such as parentheses ( ), brackets [ ], or braces { } that show you what to do first when simplifying or evaluating mathematical expressions. When grouping symbols are nested, work from the inside set to the outside set.

hexagon	A hexagon is a six-sided, two-dimensional, closed figure.

hundreds place	The hundreds place is third to the left of a decimal point, or third from the last digit of a whole number. For example, in the number 458.9, the 4 is in the hundreds place. In 789, the 7 is in the hundreds place.
hundredths place	The hundredths place is two digits to the right of the decimal point. For example, in the number 23.467, the 6 is in the hundredths place.
hypotenuse	The hypotenuse is the longest side of a right triangle. It is opposite the right angle. In the diagram, side c is the hypotenuse.

improper fraction	An improper fraction is a fraction in which the numerator is equal to or larger than the denominator. Therefore, an improper fraction is equal to or greater than 1 and can be simplified. Examples: $\frac{3}{3} = 1 \quad \frac{4}{3} = 1\frac{1}{3}$
inequality	An inequality is a mathematical sentence in which $<, >, \leq,$ or $\geq$ is used to compare two values that are not equal.
integers	Integers are the set of counting numbers, their opposites, and 0 (...-2, -1, 0, 1, 2...).
intersecting lines	Intersecting lines are lines that cross each other. Intersecting lines form vertical and adjacent angles. In the following picture, $\overleftrightarrow{AB}$ and $\overleftrightarrow{CD}$ are intersecting lines.

inverse operations	Inverse operations are opposite mathematical operations that undo each other. Addition and subtraction are inverse operations; multiplication and division are inverse operations.
isolate	To isolate is to get something by itself.
isosceles triangle	An isosceles triangle has two congruent (equal in length) sides. Triangle ABC is an isosceles triangle because $\overline{AB}$ and $\overline{AC}$ are equal in length.

kilo-	*Kilo* is a metric prefix that means 1,000. There are 1,000 meters in a kilometer and 1,000 grams in a kilogram.
least common denominator (LCD)	The least common denominator (LCD) is the smallest multiple that two denominators have in common. See least common multiple.
least common multiple (LCM)	The least common multiple (LCM) is the smallest number into which two or more numbers can divide. The LCM of 3 and 4 is 12 because 12 is the smallest multiple they have in common. Multiples of 3: 3, 6, 9, (12), 15 Multiples of 4: 4, 8, (12), 16
like terms	Like terms are terms that have identical variables or variable groups. $3a^2b$ and $7a^2b$ are like terms because a^2b and a^2b are identical. $3a^2b$ and $7ab^2$ are unlike terms because a^2b and ab^2 are not identical.
line	A line is straight, has no thickness, and extends forever in both directions. It is a straight angle. $\overleftrightarrow{AB}$ is a line.

linear	Linear refers to a graph or picture of a straight line.
mean	A mean is another word (synonym) for average. The mean of 2, 5, and 11 is found by adding $2 + 5 + 11$, which is 18, and then dividing 18 by 3 because there are three numbers. The mean is 6.

median	In a list of data arranged from lowest to highest or highest to lowest (ordered), the median is the middle number or the average of the middle two numbers.
meter	A meter is the base unit of length in the metric system. A meter is a little more than 3 feet long.
metric system	The metric system is a system of measurement that uses decimals and base ten. The liter, meter, and gram are the basic units of measurement in the metric system.
mile	A mile is a unit of measure for length that equals 5,280 feet or 1,760 yards.
milli-	*Milli* is the metric prefix that means 1/1,000th (or 0.001). A millimeter is 1/1,000th of 1 meter because there are 1,000 millimeters in 1 meter.
minuend	A minuend is the number from which another number is subtracted in a subtraction problem. In the example $12 - 5 = 7$, the 12 is the minuend.
mixed number	A mixed number is a value that includes an integer and a fraction. For example, $3\frac{1}{2}$ and $-5\frac{1}{12}$ are mixed numbers.
mode	The mode is the number that occurs the most often in a list of numbers. For example, in the list 2, 5, 4, 6, 6, 3, 6, 7, the mode is 6 because it shows up the most times.
monomial	A monomial is an algebraic expression with exactly one term. A monomial is a number, a variable, or a product of a number and a variable. For example, 3, 6a, and $-12cd$ are all monomials. *Mono* is a prefix that means 1.
multiple	A multiple of a number is the product of that number and another whole number. For example, some multiples of 2 are 2, 4, 6, and 8.
natural numbers	Natural numbers are the counting numbers: 1, 2, 3, etc.
negative number	A negative number is any number found to the left of 0 on a number line.

number line	A number line is a line with equal distances marked off to represent numbers.

numerator	The numerator is the number above the line in a fraction. The numerator tells you how many equal pieces of the whole you have. In the fraction $\frac{3}{4}$, the 3 is the numerator and tells you that you have 3 pieces of a whole that is divided into 4 equal pieces.
obtuse angle	An obtuse angle is an angle that measures more than 90° but less than 180°. An obtuse angle is bigger than a right angle (90°) but is smaller than a straight line (180°). ∠ABC and ∠DEF are obtuse angles.

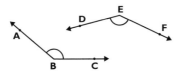

octagon	An octagon is an eight-sided, two-dimensional, closed shape. A traffic stop sign is an example of an octagon.

ones place	The ones place, also called the units place, is first to the left of a decimal point or is the last digit of a whole number. For example, in the numbers 458.9 and 618, the 8 is in the ones place.
opposites	Opposites are the positive and negative of the same number. When they are added, the sum is 0. For example, 3 and ⁻3 are opposites, and ⁻3 + 3 = 0. Opposites are the same distance from 0 on a number line.

ordered pair	An ordered pair is the coordinates (address) of a point on a coordinate (Cartesian) graph. Ordered pairs are in the form (x, y). A is located at (5, 4). C is located at (-3, -2). 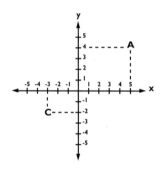
order of operations	The order of operations gives the rules for doing a problem that requires more than one operation. When doing a problem, look for these operations and do them in the following order: 1. **P**arentheses ( ) and other grouping symbols 2. **E**xponents 2^3 3. **M**ultiplication or **D**ivision from left to right 4. **A**ddition or **S**ubtraction from left to right **P**lease **E**xcuse **M**y **D**ear **A**unt **S**ally
parallel lines	Parallel lines are lines in the same plane that never intersect. $\overleftrightarrow{AB}$ and $\overleftrightarrow{CD}$ are parallel lines. Parallel lines are shown as $\overleftrightarrow{AB}//\overleftrightarrow{CD}$.
parallelogram	A parallelogram is a four-sided, two-dimensional figure that has parallel opposite sides. Figure ABCD is a parallelogram. $\overline{AB}$ is parallel to $\overline{DC}$, and $\overline{AD}$ is parallel to $\overline{BC}$.
percent	Percent is a way to represent part of a whole that has been divided into 100 equal pieces.
perimeter	The perimeter is the distance around a plane figure. The perimeter of a figure is equal to the sum of its sides.

perpendicular lines	Perpendicular lines are lines in the same plane that intersect at a right (90°) angle. $\overleftrightarrow{AB}$ and $\overleftrightarrow{CD}$ are perpendicular lines. Perpendicular lines are shown as $\overleftrightarrow{AB} \perp \overleftrightarrow{CD}$. 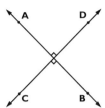
pi (or π)	Pi (π) is approximately equal to $\frac{22}{7}$ or 3.14. It is the ratio of the circumference of any circle to the diameter of the circle.
plane	A plane is a flat geometric surface. Imagine a piece of paper that extends forever.
polygon	A polygon is a two-dimensional closed shape with three or more straight sides. Examples of polygons include triangles, squares, and trapezoids.
polynomial	A polynomial is a sum of monomials. For example, 3x + 2 and 4x − 3y + 6g are polynomials.
positive number	A positive number is a number found to the right of 0 on a number line.
power	Power is another name for exponent.
prime number	A prime number is a positive whole number greater than 1 that has only two factors: itself and 1. For example, 5 is a prime number because the only way to multiply and get 5 is 1 · 5.
probability	Probability is a word meaning the chances of something occurring. The probability of flipping a coin and getting tails (or heads) is 1:2, $\frac{1}{2}$, 0.5, or 50%. Probability can be written as a ratio, fraction, decimal, or percent.
product	The product is the answer to a multiplication problem. For example, in the problem 3 · 9 = 27, the product is 27.
proportion	A proportion is a pair of equal ratios. For example, $\frac{2}{4}$ and $\frac{4}{8}$ are a proportion because they are equivalent.

protractor	A protractor is a tool used to measure and draw angles.

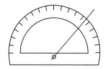

Pythagorean theorem	The Pythagorean theorem is a formula used to determine the lengths of the legs of a right triangle. The sum of the squares of the sides equals the square of the hypotenuse. The formula is $a^2 + b^2 = c^2$.

quadrant	A quadrant is one of four sections on a coordinate (Cartesian) graph where there is an x-axis (horizontal) and a y-axis (vertical). In the following picture, I, II, III, and IV are the quadrants.

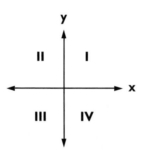

quotient	The quotient is the answer to a division problem. For example, in the problem $12 \div 2 = 6$, 6 is the quotient.
radius	The radius is a line segment from the center point of a circle to any point on the circle. It is half the length of the diameter. In the following picture, $\overline{AB}$ is the radius.

range	The range is the difference between the greatest and least values in a set of data. For example, the range of 8 to 20 is 12.

ratio	A ratio is a way of comparing a pair of numbers. Ratios appear in three forms. To compare the numbers 3 and 4, the three forms are $\frac{3}{4}$, 3 to 4, and 3:4.
ray	A ray is a line that starts at a specific point and goes on forever. $\overrightarrow{AB}$ is a ray.

real number	A real number is any integer, decimal, or fraction.
reciprocal	A reciprocal is the inverse (flip) of a fraction. For example, the reciprocal of $\frac{1}{2}$ is $\frac{2}{1}$. The product of any number and its reciprocal is always 1. Reciprocals are used to divide fractions.
rectangle	A rectangle is a four-sided, two-dimensional, closed figure that has four right (90°) angles and whose opposite sides are equal. A square is a special rectangle that has four equal sides.

remainder	A remainder is the amount left over in a division problem. A decimal point is never added to the quotient when remainders are found.
right angle	A right angle is an angle that measures exactly 90°. A right angle forms a letter L. A little square is put in the corner of a right angle to show that it is a right angle.

L ⌵ ⟩

rounding	Rounding is a way of approximating a number to a given place value. When the number 48,745 is rounded to the nearest thousands place, the approximation is 49,000 because 48,745 is closer to 49,000 than to 48,000.
segment	A segment is a piece of a line that has definite beginning and ending points. $\overline{CD}$ is a line segment.

sequence (noun)	The sequence of a set of items is the order in which the items are arranged. When people put items in a sequence, they often arrange, or order, them alphabetically or numerically.

simplify	To simplify is to combine like terms and put an answer in its lowest form.
sphere	A sphere is a three-dimensional circle. A globe and a ball are examples of spheres.
square	A square is a four-sided, two-dimensional, closed figure with four right (90°) angles and all sides equal in length.
squared	Squared, in terms of exponents, refers to an exponent (power) of 2. 3^2 is stated three squared, meaning 3 · 3 or 9.
square root	The square root of a positive number is the number that is multiplied by itself to get that positive number. For example, the square root of 9 is 3 because 3 · 3 = 9.
statistics	Statistics is the study of collecting, investigating, and displaying data.
straight angle	A straight angle is an angle that measures 180°. A straight angle is a line. ∠ABC is a straight angle.
substitute (verb)	To substitute means to replace a variable or symbol with a known numerical value. For example, if a = 4 and b = 2, then ab (or a · b) = 4 · 2 = 8.
subtrahend	The subtrahend is the number subtracted from another number. In the problem 12 − 7 = 5, the 7 is the subtrahend.
sum	The sum is the answer to an addition problem. For example, in the problem 4 + 6 = 10, 10 is the sum.
supplementary angles	Supplementary angles are angles that when added equal 180°. When combined, supplementary angles form a straight angle, or a line.
symbol	A symbol is a picture or shape that stands for an operation, a constant value, or a word or words. For example, + is the symbol for plus or addition, and π is the symbol for pi (3.14).

ten-thousands place	The ten-thousands place is fifth to the left of a decimal point or the fifth from the last digit of a whole number. For example, in the number 12,458.9, the 1 is in the ten-thousands place. The 3 in 34,798 is in the ten-thousands place.
ten-thousandths place	The ten-thousandths place is fourth to the right of a decimal point. For example, in the number 23.46789, the 8 is in the ten-thousandths place.
tens place	The tens place is second to the left of a decimal point or the second from the last digit of a whole number. For example, in the number 12,458.9, the 5 is in the tens place. The 8 in 34,789 is in the tens place.
tenths place	The tenths place is first digit to the right of a decimal point. For example, in the number 23.46, the 4 is in the tenths place.
term	A term is part of an expression. Terms are separated by addition and subtraction symbols. In the expression $4y - 2 + y$, the terms are $4y$, $^-2$, and y.
thousands place	The thousands place is fourth to the left of a decimal point or the fourth from the last digit of a whole number. For example, in the number 12,458.9, the 2 is in the thousands place. The 4 in 34,789 is in the thousands place.
thousandths place	The thousandths place is third digit to the right of a decimal point. For example, in the number 23.4678, the 7 is in the thousandths place.
trapezoid	A trapezoid is a four-sided, two-dimensional, closed figure with only one set of parallel lines. The two parallel lines are the bases. Figure ABCD is a trapezoid. $\overline{AB}$ and $\overline{CD}$ are parallel.
triangle	A triangle is a three-sided, two-dimensional, closed figure. 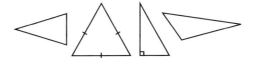

trinomial	A trinomial is a polynomial with exactly three unlike terms. An example of a trinomial is 3x + 2y − 4z. The different variables—x, y, and z—make the three terms unlike. Another example is 4 + 2d − 2d². A whole number (4) is unlike a term with a variable (2d), and a term with a variable (2d) is unlike a term with the variable with an exponent (2d²). *Tri* is a prefix that means 3.
triple	To triple a number or term means to multiply the number or term by 3. To triple 4, multiply 4 by 3, or 4 × 3, which is 12. Tripling the term 2x gives 2x • 3, which equals 6x.
unit	A unit is an amount or quantity used as a label of measurement. For example, cm, km, in², and cm³ are units used for measuring and labeling.
unlike terms	Unlike terms are terms that do not have identical variables or variable groups. For example, 3a and 4b are unlike terms because the variables are different. 3x²y and 6xy² are unlike terms because the exponents of the variables are different.
value	A value is an assigned amount. "Find the value" means to solve or to find the amount.
variable	A variable is a symbol used to take the place of an unknown number. For example, in the problem x − 7 = 3, x is the variable.
variable term	A variable term is a mathematical term with at least one variable. For example, 4x, 10ab, and 6a²b are all variable terms.
vertical angles	Vertical angles are two angles formed by intersecting lines that are across from each other. In the following picture, ∠1 and ∠2 are vertical angles. ∠3 and ∠4 are also vertical angles.

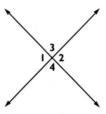

volume	Volume is the amount of space inside a three-dimensional object. Volume is labeled with cubed units.

whole numbers	Whole numbers are the counting numbers and 0 (0, 1, 2, 3…).
x-axis	The x-axis is the horizontal number line of a coordinate (Cartesian) graph.

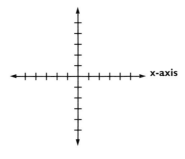

y-axis	The y-axis is the vertical number line of a coordinate (Cartesian) graph.

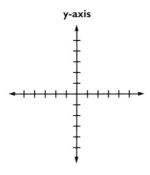

zero (0)	Zero separates the positive numbers and the negative numbers on a number line.

Algebra Ready — Scope and Sequence—Part 1   S

Lesson 1

OBJECTIVES

1. Perform the order of operations: × ÷ and + − only.
2. Practice addition, subtraction, multiplication, and division.

NCTM Standards 2000: 1 Numbers & Operations, 2 Algebra, 6 Problem Solving, 8 Communications, 9 Connections, 10 Representation

Lesson 2

OBJECTIVES

1. Perform the order of operations: exponents, × ÷, and + − .
2. Practice addition, subtraction, multiplication, and division.

NCTM Standards 2000: 1 Numbers & Operations, 2 Algebra, 6 Problem Solving, 8 Communications, 9 Connections, 10 Representation

Lesson 3

OBJECTIVE

Perform the order of operations: parentheses and other grouping symbols (brackets and braces), exponents, × ÷, and + −.

NCTM Standards 2000: 1 Numbers & Operations, 2 Algebra, 6 Problem Solving, 8 Communications, 9 Connections, 10 Representation

Lesson 4

OBJECTIVE

Review the order of operations: parentheses and other grouping symbols (brackets and braces), exponents, × ÷, and + −.

Please **E**xcuse **M**y **D**ear **A**unt **S**ally

NCTM Standards 2000: 1 Numbers & Operations, 2 Algebra, 6 Problem Solving, 8 Communications, 9 Connections, 10 Representation

Lesson 5

OBJECTIVE

Evaluate formulas, specifically the area of a circle. $A = \pi r^2$

NCTM Standards 2000: 1 Numbers & Operations, 2 Algebra, 3 Geometry, 4 Measurement, 6 Problem Solving, 8 Communications, 9 Connections, 10 Representation

Lesson 6

OBJECTIVES

1. Combine integers using the number line.
2. Apply the first three signed number rules.

NCTM Standards 2000: 1 Numbers & Operations, 2 Algebra, 6 Problem Solving, 8 Communications, 9 Connections, 10 Representation

Lesson 7

OBJECTIVES

1. Apply signed number rules 1 through 6.
2. Review signed number rules 1 through 3.

NCTM Standards 2000: 1 Numbers & Operations, 2 Algebra, 3 Geometry, 4 Measurement, 6 Problem Solving, 8 Communications, 9 Connections, 10 Representation

Lesson 8

OBJECTIVES

1. Apply the six signed number rules.
2. Review the order of operations.

NCTM Standards 2000: 1 Numbers & Operations, 2 Algebra, 6 Problem Solving, 8 Communications, 9 Connections, 10 Representation

Lesson 9

OBJECTIVES

1. Apply the six signed number rules including eliminating a double sign.
2. Review the order of operations.

NCTM Standards 2000: 1 Numbers & Operations, 2 Algebra, 4 Measurement, 5 Data Analysis & Probability, 6 Problem Solving, 8 Communications, 9 Connections, 10 Representation

Lesson 10

OBJECTIVES

1. Review the order of operations using several types of grouping symbols.
2. Practice the six signed number rules.

NCTM Standards 2000: 1 Numbers & Operations, 2 Algebra, 6 Problem Solving, 8 Communications, 9 Connections, 10 Representation

Lesson 11

OBJECTIVES

1. Review the order of operations and the signed number rules.
2. Practice addition, subtraction, multiplication, division, and exponents.

NCTM Standards 2000: 1 Numbers & Operations, 2 Algebra, 4 Measurement, 6 Problem Solving, 8 Communications, 9 Connections, 10 Representation

Lesson 12

OBJECTIVE

Solve addition and subtraction equations with variables.

NCTM Standards 2000: 1 Numbers & Operations, 2 Algebra, 6 Problem Solving, 7 Reasoning & Proof, 8 Communications, 9 Connections, 10 Representation

Lesson 13

OBJECTIVES

1. Solve multiplication equations with variables.
2. Review addition and subtraction equations.
3. Convert improper fractions.

NCTM Standards 2000: 1 Numbers & Operations, 2 Algebra, 6 Problem Solving, 7 Reasoning & Proof, 8 Communications, 9 Connections, 10 Representation

Lesson 14

OBJECTIVES

1. Solve two-step equations.
2. Reduce fractions.
3. Practice addition, subtraction, multiplication, and division with integers.

NCTM Standards 2000: 1 Numbers & Operations, 2 Algebra, 6 Problem Solving, 7 Reasoning & Proof, 8 Communications, 9 Connections, 10 Representation

Lesson 15

OBJECTIVES

1. Review two-step equations.
2. Practice addition, subtraction, multiplication, and division with integers and fractions.

NCTM Standards 2000: 1 Numbers & Operations, 2 Algebra, 5 Data Analysis & Probability, 6 Problem Solving, 7 Reasoning & Proof, 8 Communications, 9 Connections, 10 Representation

Lesson 16

OBJECTIVES

1. Simplify expressions by combining like terms.
2. Add and subtract with integers.

NCTM Standards 2000: 1 Numbers & Operations, 2 Algebra, 5 Data Analysis & Probability, 6 Problem Solving, 8 Communications, 9 Connections, 10 Representation

Lesson 17

OBJECTIVES

1. Simplify expressions by combining like terms.
2. Add and subtract with integers.

NCTM Standards 2000: 1 Numbers & Operations, 2 Algebra, 6 Problem Solving, 8 Communications, 9 Connections, 10 Representation

Lesson 18

OBJECTIVES

1. Solve two-step equations with variable terms on both sides.
2. Practice the four basic operations with integers and fractions.

NCTM Standards 2000: 1 Numbers & Operations, 2 Algebra, 3 Geometry, 4 Measurement, 6 Problem Solving, 8 Communications, 9 Connections, 10 Representation

Algebra Ready   Scope and Sequence—Part 2   S

Lesson 19

OBJECTIVES
1. Solve multi-step equations with variable terms on both sides.
2. Practice the four basic operations with integers.

NCTM Standards 2000: 1 Numbers & Operations, 2 Algebra, 6 Problem Solving, 8 Communications, 9 Connections, 10 Representation

Lesson 20

OBJECTIVES
1. Review multi-step equations with variable terms on both sides.
2. Practice the four basic operations with integers and fractions.

NCTM Standards 2000: 1 Numbers & Operations, 2 Algebra, 4 Measurement, 5 Data Analysis & Probability, 6 Problem Solving, 8 Communications, 9 Connections, 10 Representation

Lesson 21

OBJECTIVES
1. Evaluate formulas for geometric figures using diagrams and substitution.
2. Multiply and divide whole numbers, fractions, and decimals.

NCTM Standards 2000: 1 Numbers & Operations, 2 Algebra, 3 Geometry, 6 Problem Solving, 8 Communications, 9 Connections, 10 Representation

Lesson 22

OBJECTIVES
1. Review two-step and multi-step equations.
2. Practice the four basic operations with integers and fractions.

NCTM Standards 2000: 1 Numbers & Operations, 2 Algebra, 6 Problem Solving, 8 Communications, 9 Connections, 10 Representation

Lesson 23

OBJECTIVE
Solve inequalities.

NCTM Standards 2000: 1 Numbers & Operations, 2 Algebra, 6 Problem Solving, 7 Reasoning & Proof, 8 Communications, 9 Connections, 10 Representation

Lesson 24

OBJECTIVES

1. Solve equations that have fractions.
2. Practice working with fractions.

NCTM Standards 2000: 1 Numbers & Operations, 2 Algebra, 6 Problem Solving, 8 Communications, 9 Connections, 10 Representation

Lesson 25

OBJECTIVES

1. Solve equations with decimals.
2. Practice the four basic operations with decimals.

NCTM Standards 2000: 1 Numbers & Operations, 2 Algebra, 6 Problem Solving, 7 Reasoning & Proof, 8 Communications, 9 Connections, 10 Representation

Lesson 26

OBJECTIVE

Practice solving fraction and decimal equations.

NCTM Standards 2000: 1 Numbers & Operations, 2 Algebra, 6 Problem Solving, 8 Communications, 9 Connections, 10 Representation

Lesson 27

OBJECTIVE

Evaluate formulas for geometric figures using data from diagrams.

NCTM Standards 2000: 1 Numbers & Operations, 2 Algebra, 3 Geometry, 4 Measurement, 5 Data Analysis & Probability, 6 Problem Solving, 8 Communications, 9 Connections, 10 Representation

Lesson 28

OBJECTIVE

Simplify monomials using multiplication.

NCTM Standards 2000: 1 Numbers & Operations, 2 Algebra, 6 Problem Solving, 8 Communications, 9 Connections, 10 Representation

Lesson 29

OBJECTIVES

1. Simplify algebraic expressions using the distributive property.
2. Use the six signed number rules to simplify algebraic expressions.

NCTM Standards 2000: 1 Numbers & Operations, 2 Algebra, 6 Problem Solving, 8 Communications, 9 Connections, 10 Representation

Lesson 30

OBJECTIVES

1. Simplify more complex algebraic expressions using the distributive property.
2. Apply the six rules of signed numbers to simplify algebraic expressions.

NCTM Standards 2000: 1 Numbers & Operations, 2 Algebra, 6 Problem Solving, 8 Communications, 9 Connections, 10 Representation

Lesson 31

OBJECTIVE

Simplify algebraic expressions using the distributive property and the signed number rules to distribute a plus (+) sign through parentheses.

NCTM Standards 2000: 1 Numbers & Operations, 2 Algebra, 3 Geometry, 4 Measurement, 6 Problem Solving, 8 Communications, 9 Connections, 10 Representation

Lesson 32

OBJECTIVE

Simplify algebraic expressions using the distributive property and the signed number rules to distribute a minus (–) sign through parentheses.

NCTM Standards 2000: 1 Numbers & Operations, 2 Algebra, 6 Problem Solving, 8 Communications, 9 Connections, 10 Representation

Lesson 33

OBJECTIVE

Solve equations using the distributive property and combining like terms.

NCTM Standards 2000: 1 Numbers & Operations, 2 Algebra, 6 Problem Solving, 7 Reasoning & Proof, 8 Communications, 9 Connections, 10 Representation

Lesson 34

OBJECTIVES

1. Solve multi-step equations with variable terms on both sides.
2. Practice using the order of operations, the signed number rules, the distributive property, and inverse operations.

NCTM Standards 2000: 1 Numbers & Operations, 2 Algebra, 6 Problem Solving, 8 Communications, 9 Connections, 10 Representation

Lesson 35

OBJECTIVES

1. Practice solving multi-step equations with variable terms on both sides.
2. Practice using the order of operations, the signed number rules, the distributive property, and inverse operations.

NCTM Standards 2000: 1 Numbers & Operations, 2 Algebra, 6 Problem Solving, 8 Communications, 9 Connections, 10 Representation

Lesson 36

OBJECTIVE

Apply algebra to real-world problem solving.

NCTM Standards 2000: 1 Numbers & Operations, 2 Algebra, 3 Geometry, 4 Measurement, 5 Data Analysis & Probability, 6 Problem Solving, 8 Communications, 9 Connections, 10 Representation

Note: The standards are reprinted with permission from *Principles and Standards for School Mathematics*, copyright 2000, by the National Council of Teachers of Mathematics.

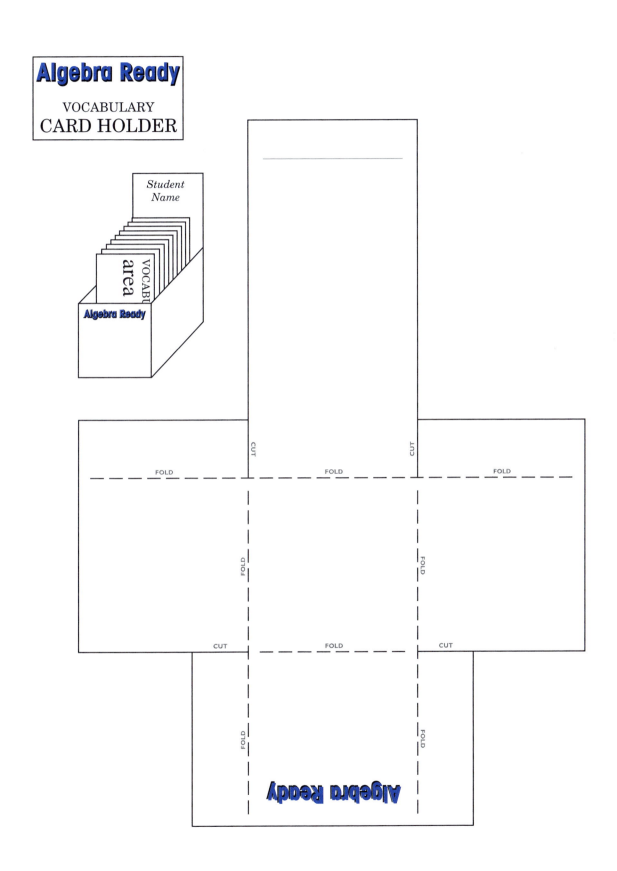

(SCHOOL LOGO)

This certificate is awarded to

_____

for successfully completing

Algebra Ready

PART 1

_____
Algebra Ready Math Teacher

_____
Principal

_____
Date